BIOPHOTONICS

Optical Science and Engineering
for the 21st Century

BIOPHOTONICS

Optical Science and Engineering for the 21st Century

Edited by

Xun Shen
Institute of Biophysics
Chinese Academy of Sciences
Beijing, China

and

Roeland Van Wijk
Station Hombroich
International Institute of Biophysics
Neuss, Germany

 Springer

Additional material to this book can be downloaded from http://extras.springer.com

ISBN-10: 0-387-24995-8
ISBN-13: 978-0387-24995-7
eISBN: 0-387-24996-6

©2005 Springer Science+Business Media, Inc.

All rights reserved. This work may not be translated or copied in whole or in part without the written permission of the publisher (Springer Science+Business Media, Inc., 233 Spring Street, New York, NY 10013, USA), except for brief excerpts in connection with reviews or scholarly analysis. Use in connection with any form of information storage and retrieval, electronic adaptation, computer software, or by similar or dissimilar methodology now known or hereafter developed is forbidden.

The use in this publication of trade names, trademarks, service marks and similar terms, even if they are not identified as such, is not to be taken as an expression of opinion as to whether or not they are subject to proprietary rights.

Printed in the United States of America

9 8 7 6 5 4 3 2 1   SPIN 11393429

springeronline.com

PREFACE

Biophotonics deals with interactions between photons and biological matter. It is an exciting frontier that involves a fusion of photonics and biology. Biophotonics is the science of generating and harnessing light (photons) to image, detect and manipulate biological materials. It offers great hope for the early detection of diseases and for new modalities of light-guided and light-activated therapies. It also provides powerful tools for studying molecular events, such as gene expression, protein-protein interaction, spatial and temporal distribution of the molecules of biological interest, and many chemico-physical processes in living cells and living organisms. Fluorescence, scattering and penetrating light are frequently used to detect and image the biological systems at molecular, cellular and organismic levels. The light generated by metabolic processes in living organisms also provides a good optical means to reflect the structure and function of the living cells and organisms, which leads to a special aspect of biophotonics called "Biophoton research". Either biophotonics or biophoton research creates many opportunities for chemists, physicists, biologists, engineers, medical doctors and healthcare professionals. Also, educating biomedical personnel and new generations of researchers in biophotonics is of the utmost importance to keep up with the increasing worldwide demands.

On October 12-16 2003, scientists from 12 nations met in Beijing in order to present and to discuss the most recent results in the interdisciplinary field of biophotonics and biophoton research. Profound discussions were devoted to the new spectroscopic techniques in microscopic imaging and optical tomography that allow determination of the structures and functions in cells and tissues. Besides problems of basic research in this field of biophotonics, various applications of these new optical technologies in non-invasive or minimally invasive optical imaging, monitoring, and sensing of complex systems such as tissues at the cellular level and cells at the subcellular level have been presented. Scientists working in the particular field of biophoton research presented new and exciting experimental results on spontaneous photon emission from living organisms. They discussed the probable light sources within the cells, the possible coherence of the photon field within the organism, and its bio-communicative aspects.

The field of biophotonics and biophoton research as covered in this book is an important step forward in our understanding of the essence of biology, which is composite and complex. Biology studies organisms: objects, which are complicated enough to live. These objects cannot be understood by reducing life to a simple summation of singular properties of many molecular components. In this respect, biophotonics and biophoton research offer a possibility to rise above biochemical reductionism approach of molecular biology and study with success life within the concept of "quantum biology". In the 1930's, Pascual Jordan, one of the founders of quantum theory has already proposed the concept of "quantum biology". The name is recognized in the famous "Cold Spring Harbor Symposia on Quantum Biology", which were originally aimed at the new understanding of biology with the new developments in physics and chemistry. However, the consequences of quantum physics have not been made in biology, even not in the "Cold Spring Harbor Symposia on Quantum Biology", which were strongly involved in the birth of molecular biology. In fact, at that time the reductionism paradigm, which assumes that systems can be understood by simply accounting for the properties of their parts, was too overwhelming in biology. In particular during the time that there was no clear theory of the gene's inner working, the Watson–Crick discovery of the double helix led to a successful development of the paradigm of the gene as ultimate control agent. It has the additional effect of diminishing the concept of the organism in experimental biology. In the mid-20th century, the life sciences in universities throughout the world still maintained strong programs in organismal biology, but in the 1980's and 1990's, in most universities, especially in Western Europe and United States, these activities were segregated into historical departments. A profound shift occurred in our perception of the world (from organisms to gene machines) in which we learned more and more about molecules and molecular reactions and less about life. In fact, the field of biophotonics and biophoton research has important consequences for the re-discovery and implementation of the quantum physics concept in biology.

The new developments in spectroscopic techniques might allow the detection of functioning of multi-component complexes constructed from many proteins with their rules not written in DNA. They can lead to insight into the fundamental interconnectedness within the organism. Many examples ranging from molecular to cellular level can be listed as examples of patterns of an emergent complexity with rules that can now be studied. Pattern formation and phase transitions in complex cytoskeletal protein structures in connection with cell movement and dynamic cell structure are for example processes that belong to this category. Another example is the synchronization and emergence of oscillations.

The role of biophoton research in quantum biology is also extensively illustrated in this book. In particular this approach, a new type of biophysics, will focus on holistic aspects of the organism. It attempts to provide a new vision able to synthesize the wealth of molecular details accumulated by molecular biologists. Biophoton research is focused on understanding life processes by reading the language of the ultraweak spontaneous emission from living tissues and cells (proper biophoton emission). However, in close connection has been considered the language of cell's light production following excitation, the light-induced process of delayed emission of light which can be detected for a long term from biological systems after excitation. The progress in this area includes new developments in technologies for 2- or 3-dimensionally imaging biophotons, as well as for analyzing the properties of this ultraweak emission and its biological significance. One of the most essential questions concerns the question of coherence and incoherence of biophotons.

In the text we strive to present the story of biophotonics and biophoton research in a

PREFACE						vii

clear and integrated manner. Although the chapters of this book can be read independently of one another, they are arranged in a logical sequence. The book is divided in two sections: Biophotonics (including 8 chapters) and Biophotons (including 8 chapters)

In "Biophotonics" section, Chapter 1 describes the newly developed method for studying biochemical reactions in the cell interior using Fluctuation Correlation Spectroscopy (FCS) in combination of two-photon fluorescence excitation. Consider a very small volume (femtoliter volume) within a cell and the fluorescent dye-labeled protein molecules of biological interest in the cell, the fluorescence fluctuation arises because of the chemical reaction that changes the fluorescence properties of the dye and because the bound molecules could enter and leave the volume of excitation due to diffusion of the molecule. Thus, the statistical analysis of fluctuations of the fluorescence signal provides a powerful tool for the study of chemical reactions both in solutions and in the interior of cells.

Chapter 2 describes a new light microscopy called Evanescent Wave Microscopy, in which a novel objective lens that has an ultrahigh numerical aperture of 1.65 is used. This new light microscope can be used to study dynamics of the cell membrane by observation of fluorescent objects of biological specimens under the illumination of evanescent light. In this chapter, the studies on ion channel, protein kinase C, dynamin, inositol trisphosphate and exocytosis using evanescent wave microscopy are discussed.

Chapter 3 and 4 both describe the novel optical technologies, which use different colored fluorescent proteins, the fluorescent protein-gene fusion technique and fluorescence resonance energy transfer (FRET), and their applications in studying the molecular processes in single living cell. The former demonstrates how these innovative optical technologies are used to study signaling mechanisms in programmed cell death, and the latter demonstrates how these innovative optical technologies are used to study protein-protein interaction in single living cell using the interaction between small heat shock protein and p38 MAP kinase as an example. Some imaging technique such as fluorescence lifetime imaging is also discussed. Both chapters may well demonstrate that the biophotonics is probably the best solution for understanding cell function by integrating molecular activities within the living cells. To integrate molecular activities within a single living cell has been a big challenge for modern biology.

Optical coherence tomography (OCT) is a recently developed imaging modality based on coherence-domain optical technology. The high spatial resolution of OCT enables noninvasive *in vivo* "optical biopsy" and provides immediate and localized diagnostic information. Chapter 5 reviews the principle of OCT and Functional OCT (F-OCT) and highlights some of the results obtained in the OCT Laboratory at the Beckman Laser Institute.

Chapter 6 describes a modified laser speckle imaging (LSI) technique. LSI is kind of intrinsic optical imaging and based on the temporal statistics of a time-integrated speckle. The laser speckle is an interference pattern produced by the light reflected or scattered from different parts of the illuminated tissue area (in author's investigation it is the cerebral cortex of rat). It has been demonstrated that the motion information of the scattering particles could be determined by integrating the intensity fluctuations in a speckle pattern over a finite time. In areas of higher blood flow the speckle intensity fluctuations are more rapid and when integrated over a finite time, these areas show increased blurring of the speckle pattern. In this chapter, both the principle of LSI and its application for monitoring the spatio-temporal characteristics of cerebral blood flow in brain are discussed.

Except for fluorescence probes that are frequently used for imaging the molecules of biological interest in cell and tissue, chemiluminescence probes can also be used for the same purpose. Chapter 7 describes a novel method of photodynamic and sonodynamic diagnosis of cancer by using chemiluminescence probe. The method is based on two basic principles: (1) photosensitizer, such as heamatoporphorin derivatives, is preferentially accumulated in cancer tissues, and (2) the light- or ultrasound-induced reaction of the photosensitizer with molecular oxygen yields reactive oxygen species that further react with chemiluminscence probe (such as *Cypridina* luciferin analogue) to give rise to photon emission from the photosensitizer-bearing tissue.

Chapter 8 introduced a very useful method for characterizing molecular chaperones that help protein folding and refolding. The denatured luciferase is used as polypeptide to be refolded, and the luciferase-catalyzed bioluminescence is used as a measure for evaluating the function of the studied chaperone in helping protein refolding.

In "Biophotons" section, chapter 9 and 10 cover elementary principles and basic biophysics. It can serve as an introduction for those who have not studied the aspects of coherence and biophoton field. It discusses the concept of "coherent states" which transcends the classical concept of coherence. Coherent states are not just characterized by the ability for interference ("coherence of the second order"), but fulfill the ideal relations of a coherence of an arbitrarily high order. The importance of coherent states for biological systems is discussed: they enable them to optimize themselves concerning organization, information quality, pattern recognition, etc. Squeezed states as a more general concept than coherent states are also discussed, the latter being considered as special cases of the former.

Chapter 11 discusses the specification of the source of energy, which continuously pumps the biophotonic field. It deals with ultraweak photon emission as chemiluminescence resulting from relaxation of electronically excited states generated in reactions with the participation of reactive oxygen species. They have been until recently considered as by-products of biochemical processes, a view that considers ultraweak photon emission as irrelevant to the performing of vital functions. The gradual erosion of this point of view and the gradual increase in research devoted to the participation of reactive oxygen species in the regulation of a wide spectrum of biochemical and physiological functions is discussed.

Chapter 12, 13, 14, 15 and 16 cover modern technologies for the determination and analysis of biophotons and several studies (chiefly based on the imaging of biophotons) for biological and medical applications aimed at diagnostic use. Although the development of the photomultiplier tube in 1950s has allowed the fundamental photon emission phenomena to be revealed, sophisticated techniques for analyzing the faint emissions have been developed for further progress in biophoton applications. Examples are the moveable photomultiplier tube in ultralow-noise dark room, which allows the recording of large surfaces (for instance human body studies), and the image system for biophoton emission consisting of two-dimensional photon counting tubes, and CCD camera systems. In terms of feasibility studies for biomedical applications, experimental results obtained from the measurement of plants and mammals clarify the relationship between biophotons and pathophysiological responses. The studies include the response of plants and animals to stress, and the biophoton emission from the brain associated with neuronal activity, and the biophoton studies of a human body. A new highly sensitive method for light-induced delayed ultraweak luminescence is discussed. The utilization of delayed luminescence method is illustrated in studies on animal cells and plants. The final chapter in this book, focusing on plant defense mechanisms, illustrates the recording of photon emission utilizing special chemiluminescence probes as sensitizers.

Included with the book is a CD containing electronic files of the color figures reproduced in black and white in the text.

We are facing a fast-expanding field of biophotonics, where this exciting topic of digitized imaging techniques with modern optics will become the optical science and engineering for the 21st century. We also facing another exciting field of biophoton, where the photons emitted from almost all living organisms may become a focal point of interdisciplinary scientific research in revealing, probably a basic, up to now widely unknown channel of communication within and between cells, stimulating thus a new scientific approach to understanding the nature of life.

Xun Shen
Professor, Institute of Biophysics
Chinese Academy of Sciences
Beijing, China

Roeland Van Wijk
Professor, International Institute of Biophysics
Neuss, Germany

CONTENTS

1. FLUCTUATION CORRELATION SPECTROSCOPY IN CELLS: DETERMINATION OF MOLECULAR AGGREGATION

E. Gratton, S. Breusegem, N. Barry, Q. Ruan, and J. Eid

1.	INTRODUCTION	1
2.	METHODS TO PRODUCE A CONFOCAL OR SMALL VOLUME	2
3.	ADVANTAGES OF TWO-PHOTON EXCITATION	3
4.	FCS: TIME AND AMPLITUDE ANALYSIS	3
5.	FLUCTUATIONS IN CELLS: PROTEIN-MEMBRANE INTERACTIONS	7
6.	CROSS-CORRELATION METHODS	9
7.	CROSS-CORRELATION AND MOLECULAR DYNAMICS	11
8.	CONCLUSIONS	13
9.	ACKNOWLEDGEMENTS	13
10.	REFERENCES	13

2. DYNAMICS OF THE CELL MEMBRANE OBSERVED UNDER THE EVANESCENT WAVE MICROSCOPE AND THE CONFOCAL MICROSCOPE

Susumu Terakawa, Takashi Sakurai, Takashi Tsuboi, Yoshihiko Wakazono, Jun-Ping Zhou, and Seiji Yamamoto

1.	INTRODUCTION	15
2.	ULTRA HIGH NA OBJECTIVE LENS	15
3.	OBSERVATIONS UNDER EVANESCENT WAVE ILLUMINATION	18
	3.1. Calcium Indicator Dye	18
	3.2. Ion Channel	19
	3.3. Protein Kinase-C	19

	3.4. Dynamin	20
	3.5. IP$_3$	21
	3.6. Exocytosis	21
4.	DISCUSSION	22
5.	ACKNOWLEDGEMENTS	23
6.	REFERENCES	23

3. USING GFP AND FRET TECHNOLOGIES FOR STUDYING SIGNALING MECHANISMS OF APOPTOSIS IN A SINGLE LIVING CELL

Donald C. Chang1, Liying Zhou, and Kathy Q. Luo

1.	INTRODUCTION	25
2.	USING THE GFP-GENE FUSION TECHNIQUE TO STUDY THE DYNAMIC REDISTRIBUTION OF SIGNALING PROTEINS IN A SINGLE LIVING CELL	27
	2.1. General Properties of GFP	27
	2.2. Application of the GFP Technology for Biophotonic Studies of Programmed Cell Death	28
	2.3. Using the GFP Technology to Test a Specific Model of Bax Activation	31
3.	APPLICATION OF THE FRET TECHNIQUE	31
	3.1. The Principle of FRET	31
	3.2. Designing a Bio-sensor for Measuring Caspase-3 Activity Based on FRET	32
	3.3. In Vitro Characterization of Sensor C3 Using Purified Protein	33
	3.4. Application of Sensor C3 to Measure the Dynamic Activation of Caspase-3 in a Single Living Cell During Apoptosis	35
	3.5. Advantages of the FRET-based Bio-sensor	37
4.	ACKNOWLEDGEMENT	37
5.	REFERENCE	38

4. FLUORESCENCE RESONANCE ENERGY TRANSFER (FRET) STUDY ON PROTEIN-PROTEIN INTERACTION IN SINGLE LIVING CELLS

Xun Shen, Chunlei Zheng, Ziyang Lin, Yajun Yang, and Hanben Niu

1.	INTRODUCTION	39
2.	PRINCIPLE OF FLUORESCENCE RESONANCE ENERGY TRANSFER (FRET)	40
3.	FRET MICROSCOPY: METHODS FOR FRET MEASUREMENT	42
	3.1. Sensitized Acceptor Fluorescence	42
	3.2. Acceptor Photobleaching Approach	42
	3.3. Fluorescence Lifetime Imaging Microscopy (FLIM)	43

4.	STUDIES ON THE INTERACTION BETWEEN HEAT SHOCK PROTEIN 27 AND P38 MAP KINASE			43
	4.1. Small Heat Shock Protein 27 and p38 MAP Kinase			43
	4.2. Experimental procedures			44
		4.2.1.	Expression Vectors and Cell Transfection	44
		4.2.2.	Microscopy	44
	4.3. Results			45
		4.3.1.	Interaction of p38 and Hsp27 in Quiescent Cell	45
		4.3.2.	Interaction of p38 with Hsp27 in H_2O_2-stimulated Cell	46
		4.3.3.	Interaction of p38 with hsp27 in the Cell Stimulated by Arachidonic Acid	49
5.	DISCUSSION			50
6.	ACKNOWLEDGEMENTS			51
7.	REFERENCES			51

5. FUNCTIONAL OPTICAL COHERENCE TOMOGRAPHY: SIMULTANEOUS IN VIVO IMAGING OF TISSUE STRUCTURE AND PHYSIOLOGY

Zhongping Chen

1.	INTRODUCTION	53
2.	OPTICAL COHERENCE TOMOGRAPHY	54
3.	FUNCTIONAL OPTICAL COHERENCE TOMOGRAPHY	59
	3.1. Doppler OCT	59
	3.2. Polarization Sensitive OCT	63
	3.3. Second Harmonic OCT	66
4.	CONCLUSIONS	69
5.	ACKNOWLEDGMENTS	69
6.	REFERENCES	69

6. TEMPORAL CLUSTERING ANALYSIS OF CEREBRAL BLOOD FLOW ACTIVATION MAPS MEASURED BY LASER SPECKLE CONTRAST IMAGING

Qingming Luo and Zheng Wang

1.	INTRODUCTION	73
2.	PRINCIPLES OF LASER SPECKLE IMAGING	75
3.	METHODS	77
4.	RESULTS	77
5.	DISCUSSIONS	82
6.	ACKNOWLEDGMENTS	83
7.	REFERENCES	83

7. PHOTO- AND SONODYNAMIC DIAGNOSIS OF CANCER MEDIATED BY CHEMILUMINESCENCE PROBES

Da Xing and Qun Chen

1.	INTRODUCTION	85
2.	MATERIALS AND METHODS	86
	2.1. Reagents	86
	2.2. Experimental setup	87
	2.3. Preparation of tumor model	88
3.	RESULTS AND DISCUSSIONS	88
	3.1. Measurement of photosensitized chemiluminescence mediated by FCLA in model solution	88
	3.2. Effects of quenchers and D_2O on photosensitized chemiluminescence mediated by FCLA in model solution	89
	3.3. Imaging of photosensitized chemiluminescence mediated by FCLA in tumor-bearing nude mouse	90
	3.4. Measurement of sonosensitized chemiluminescence mediated by MCLA in model solution	91
	3.5. Imaging of sonosensitized chemiluminescence mediated by FCLA in tumor-bearing nude mouse	94
	3.6. The Time-dependence of FCLA retention in cells	94
4.	CONCLUSIONS	95
5.	ACKNOWLEDGMENTS	96
6.	REFERENCES	96

8. THE LUMINESCENCE ASSAYS FOR THE LUCIFERASE REFOLDING FACILITATED BY HUMAN CHAPERONE MRJ IN VITRO

Meicai Zhu, Chenggang Liu, Ying Liu, Yinjing Wang, Tao Chen, Xinhua Zhao, and Yaning Liu

1.	INTRODUCTION	99
2.	THE FUNCTION AND STRUCTURE OF MOLECULAR CHAPERONES	99
3.	LUCIFERASE AS A MOLECULAR MODEL IN CHAPERONE STUDIES	101
4.	EXPERIMENT	101
	4.1. The isolation and identification of MRJ	101
	4.2. The expression and distribution of MRJ in the cells.	101
	4.2.1. The expression of MRJ in COS-7 cell.	101
	4.2.2. The expression and distribution of MRJ in CHO cell cycle	102
	4.3. The luminescence assays for the luciferase refolding facilitated by MRJ	103
	4.3.1. The MRJ expression and protein purification.	103
	4.3.2. The denature of luciferase	103

		4.3.3.	The MRJ facilitated luciferase repatriation verified by bioluminescence	104
	5.	DISCUSSION		106
	6.	ACKNOWLEDGMENT		107
	7.	REFERENCES		107

9. ESSENTIAL DIFFERENCES BETWEEN COHERENT AND NON-COHERENT EFFECTS OF PHOTON EMISSION FROM LIVING ORGANISMS

Fritz-Albert Popp

1.	BASIC REMARKS	109
2.	CLASSICAL VERSUS QUANTUM COHERENCE	114
3.	EXPERIMENTAL SITUATION	120
4.	ACKNOWLEDGEMENTS	122
5.	REFERENCES	122

10. PARAMETERS CHARACTERIZING SPONTANEOUS BIOPHOTON SIGNAL AS A SQUEEZED STATE IN A SAMPLE OF PARMELIA.TINCTORUM

R. P. Bajpai

1.	INTRODUCTION	125
2.	PHOTO COUNT STATISTICS OF A SQUEEZED STATE	128
3.	MATERIALS AND METHOD	130
4.	RESULTS AND DISCUSSIONS	131
5.	IMPLICATIONS OF SQUEEEZED STATE	134
	5.1. Possible semi classical scenario of biophoton emission	135
	5.2. Vitality indicators and indices of vitality	138
	5.3. New vistas of reality	139
6.	REFERENCES	140

11. BIOPHOTONIC ANALYSIS OF SPONTANEOUS SELF-ORGANIZING OXIDATIVE PROCESSES IN AQUEOUS SYSTEMS

Vladimir L. Voeikov

1.	INTRODUCTION	141
2.	PATHWAYS OF OXYGEN CONSUMPTION IN LIVING ORGANISMS	142
3.	BIO-REGULATORY FUNCTIONS OF PROCESSES ACCOMPANIED WITH GENERATION OF ELECTRON	144

EXCITATION ENERGY
 3.1. Biophoton Emission from Non-diluted Human Blood.   144
 3.2. Autoregulation in Model Aqueous Systems Related to ROS   148
   Production and EEE Generation.
 3.3. Probable role of water.   152
4. GENERAL CONCLUSIONS   153
5. REFERENCES   154

12. TWO-DIMENSINAL IMAGING AND SPATIOTEMPORAL ANALYSIS OF BIOPHOTON TECHNIQUE AND APPLICATIONS FOR BIOMEDICAL IMAGING

Masaki Kobayashi

1. INTRODUCTION   155
2. TWO-DIMENSIONAL DETECTION AND ANALYSIS TECHNIQUE   156
 OF BIOPHOTONS
 2.1. Photon Counting Imaging and Spatiotemporal Analysis   156
 2.2. CCD Imaging   158
3. CHARACTERIZATION OF BIOPHOTON PHENOMENA FOR   159
 BIOLOGICAL MEASUREMENTS AND APPLICATIONS
 3.1. Plants   160
 3.2. Mammal   163
 3.3. Rat Brain   163
  3.3.1. Correlation Between Photon Emission Intensity and EEG   164
    Activity
  3.3.2. Ultraweak Photon Emission Spectra of Brain Slices   165
 3.4. Human body   168
4. CONCLUSION   169
5. ACKNOWLEDGEMENTS   169
6. REFERENCES   170

13. ULTRAWEAK PHOTON EMISSION FROM HUMAN BODY

Roeland Van Wijk and Eduard Van Wijk

1. INTRODUCTION   173
2. SPONTANEOUS PHOTON EMISSION AND DELAYED   174
 LUMINESCENCE FROM HUMAN SKIN
3. DETERMINATION OF TIME SLOTS FOR RECORDING PHOTON   175
 EMISSION
4. TOPOGRAPHICAL VARIATION OF SPONTANEOUS PHOTON   176
 EMISSION
5. EFFECTS OF COLORED FILTERS ON SPONTANEOUS VISIBLE   178
 EMISSION

6.	EFFECTS OF COLORED FILTERS ON LIGHT-INDUCED DELAYED LUMINESCENCE	181
7.	SPONTANEOUS EMISSION AND BLOOD SUPPLY	181
8.	DISCUSSION	181
9.	PERSPECTIVES	183
10.	ACKNOWLEDGEMENTS	183
11.	REFERENCES	184

14. LASER-ULTRAVIOLET-A INDUCED BIOPHOTONIC EMISSION IN CULTURED MAMMALIAN CELLS

Hugo J. Niggli, Salvatore Tudisco, Giuseppe Privitera, Lee Ann Applegate, Agata Scordino, and Franco Musumeci

1.	INTRODUCTION	185
	1.1. Overview	185
	1.2. History	185
	1.3. Biophotonic sources	186
	1.4. Ultraweak photons in cultured cells	186
2.	MATERIAL AND METHODS	187
	2.1. Cell Culturing	187
	2.2. Classical Biophotonic Measurements in Human Skin Cells	187
	2.3. Delayed Luminescence Measurements in Mammalian Cells after UVA laser induction	188
3.	Results and Discussion	189
	3.1. Biophotonic emission in cultured cells with the classical design	189
	3.2. Ultraweak photon emission in mammalian cells following irradiation with a nitrogen laser in the UVA-range	191
	3.3. Conclusion	192
4.	ACKNOWLEDGEMENTS	193
5.	REFERENCES	193

15. BIOPHOTON EMISSION AND DELAYED LUMINESCENCE OF PLANTS

Yu Yan

1.	INTRODUCTION	195
2.	MATERIALS AND METHODS	196
	2.1. Instrument of BPE and DL measurements	196
	2.2. Measurement of BPE of germinating barley seeds	196
	2.3. Measurement of DL of plant leaves	197
3.	RESULTS AND DISCUSSION	197
	3.1. BPE of germinating barley seeds	197
	3.2. DL of plant leaves	201
4.	CONCLUSION	202
5.	REFERENCE	203

16. BIOPHOTON EMISSION AND DEFENSE SYSTEMS IN PLANTS

Takahiro Makino, Kimihiko Kato, Hiroyuki Iyozumi, and Youichi Aoshima

1.	INTRODUCTION	205
	1.1. Overview of the Plant Defense Cascade System	205
	1.2. Spectral Analyses of the Photon Emission and Change of Physiological State in the Defense Response	206
	1.3. Approach to Possible Sources of Photon Emission in the Defense System	206
2.	MATERIALS AND METHODS	207
	2.1. A Photon Counting System	207
	2.2. Inoculation of Microorganisms and the Treatment of Reagents	207
	2.3. Application of 2, 4-Dichlorophenoxyacetic Acid	207
	2.4. Control of Sample Temperature	208
	2.5. Spectral Analyses	208
	2.6. Photon Emission and Enzyme Reactions In Vitro	209
	2.7. Photon Emission and Enzyme Reactions In Vivo	209
	2.7.1. Application of the Enzyme Solutions	209
	2.7.2. Application of the Substrate Solutions	209
	2.7.3. Application of the Enzyme-Reaction Inhibitors	210
3.	RESULTS	210
	3.1. Time-Dependent Analyses of Biophoton Emission	210
	3.2. Spectral Analyses of Biophoton Emission	211
	3.3. Comparative Analyses with 2, 4-D and Alternating Temperature Treatments	211
	3.4. Photon Emission and Enzyme Reactions In Vitro	215
	3.5. Photon Emission and Enzyme Reactions In Vivo	215
	3.5.1. Application of the Enzyme Solutions	215
	3.5.2. Application of the Substrate Solutions	215
	3.5.3. Application of the Enzyme-Reaction Inhibitors	215
4.	DISCUSSION	215
5.	SUMMARY	217
6.	REFERENCES	218

INDEX   219

Chapter 1

FLUCTUATION CORRELATION SPECTROSCOPY IN CELLS:
Determination of molecular aggregation

E. Gratton[1], S. Breusegem[2], N. Barry[2], Q. Ruan[3], and J. Eid[4]

INTRODUCTION

Fluorescence Correlation Spectroscopy (FCS) was first introduced by Elson, Madge and Webb (1-5) for studying the binding process between ethydium bromide and DNA. When the ethydium dye binds to DNA its fluorescence quantum yield changes by a large factor. It is essentially not fluorescent when free in solution and it becomes strongly fluorescent when bound to double strand DNA. Although the processes are very different in nature, the instrumentation used for the FCS experiment is derived from dynamic light scattering. There are however major differences between dynamic light scattering and FCS. In the FCS experiment the fluorescence fluctuation arises because of the chemical reaction that changes the fluorescence properties of the dye and because the bound ethydium molecules could enter and leave the volume of excitation due to diffusion of the molecule. In dynamic light scattering the fluctuations arise from changes in the index of refraction due to local changes in the concentration of molecules. Therefore FCS is sensitive to all chemico-physical processes that could change the fluorescence intensity in a small volume. For the fluctuation in intensity to be measurable it is crucial that the volume of observation to be small so that only a few molecules are at any instant of time in the volume of observation.

The realization of small exaction volume was a major problem hindering the use of this technique and only recently, with the introduction of confocal microscopy and two-photon microscopy FCS the generation of sub-femtoliter volumes of observation has become a routine procedure. At nanomolar concentration a femtoliter volume contains

---

[1] Enrico Gratton, Laboratory for Fluorescence Dynamics, 1110 W. Green Street, University of Illinois at Urbana-Champaign, Urbana, IL 61801; [2] Sophia Breusegem, Nicholas Barry, University of Colorado Health Sciences Center, Denver, CO 80262; [3] Qiaoqiao Ruan, Abbott Laboratories, Abbott, IL 60064; [4] John Eid, Rowland Institute at Harvard, Cambridge, MA 02142

only few molecules. As a consequence of the Poisson distribution of the occupation number, the fluctuations in fluorescent intensity of a fluorescent dye in this small volume are appreciable. There are many methods to study chemical reactions and diffusion of molecules in solutions, but when it comes to the study of these reactions in cells, the problem becomes almost insoluble. The appeal of the FCS technique is that the confocal volume can be placed anywhere in the cell without disrupting the cell membrane of perturbing the cell. Therefore, as soon as two-photon microscopy was introduced by Denk, Strickler and Webb (6) we started to develop methods to study reactions in the cell interior using FCS in combination of two-photon fluorescence excitation (7,8). The necessary ingredients for FCS to work are methods to excite a small volume and very high sensitivity and dynamic range. The recent developments in microscopy, in new ultra-sensitive detectors and fast computers have made FCS a relatively simple to use technique. Commercial instruments are now available from several manufactures and the number of publications using the FCS technique is increasing very much.

Methods to produce a confocal or small volume

There are several possibilities for producing a relatively small volume for fluorescence excitation. These methods can be classified in two broad classes: methods that are limited by the wavelength of light and methods that are limited by construction of restricted volumes. Methods that are not limited by the wavelength of light are based on nanolithography, local field enhancements and near-field effects. Using these methods very small volumes can be achieved, on the order of 100 nm or smaller in size. However, these methods are not applicable to study the interior of cells. The methods commonly employed in cell studies are all limited by the wavelength of light and they are based on the following principles: confocal volume limited by the size of pinholes, multi-photon effects limited by the order of photon excitation, second harmonic generation, similar in volume to two photon excitation, stimulated emission and four-way mixing (Coherent Anti-Stokes Raman Scattering). In our lab, we have developed methods based on two-photon excitation. The excitation volume characteristic of two photon excitation has been approximated by a Gaussian-Lorentzian shape. The effective volume of excitation is about 0.1 fL, limited by the wavelength of the light used and by the numerical aperture of the objective. The point-spread-function (PSF) for one-photon excitation was modeled by the following expression (7):

$$I_{GL}(r,z) = I_0 \frac{2}{\pi} e^{-2\left(\frac{r^2}{w(z)^2}\right)} \left(\frac{w_0^2}{w(z)^2}\right)$$

where w(z) is related to the wavelength of the excitation source, λ, and the numerical aperture (NA) of the objective in the following manner:

$$w(z) = w_0 \left(1 + \left(\frac{z}{z_r}\right)^2\right)^{1/2}, \quad z_r = \frac{\pi w_0^2}{\lambda}, \text{ and } w_0 \approx \frac{1.22\lambda}{NA}.$$

w_0 is the diffraction limited $1/e^2$ beam waist (7).

Table 1. Orders of magnitude of number of molecules and diffusion time (for 1 μM solution, small molecule, water) in different volume of excitation.

Volume	Device	Size(μm)	Molecules	Time (s)
milliliter	cuvette	10000	6×10^{14}	10^4
microliter	plate well	1000	6×10^{11}	10^2
nanoliter	microfabrication	100	6×10^8	1
picoliter	typical cell	10	6×10^5	10^{-2}
femtoliter	confocal volume	1	6×10^2	10^{-4}
attoliter	nanofabrication	0.1	6×10^{-1}	10^{-6}

In Table 1, we report typical volumes that can be obtained using different techniques, the number of molecules in the volume (at a given bulk concentration) and the average time that a small molecule (in water at room temperature) will take to transit through that volume by random diffusion.

Advantages of two-photon excitation

There are several distinct advantages of the two-photon excitation method for the study of the cellular environment essentially due to the relatively low photo-toxicity of the near-ir radiation. Of course, the intrinsic two-photon excitation sectioning effects makes it possible to place the volume of illumination virtually everywhere in the cell body. For tissue work the penetration depth of two-photon excitation could be particularly useful. From the spectroscopic point of view, there is large separation between excitation and emission and virtually no second-order Raman effect. The high degree of polarization and the wavelength dependence of two-photon excitation which for several dyes extend over a large spectral region, can be exploited for specific applications based on light polarization.

FCS: time and amplitude analysis

In a typical fluctuation experiment a small volume is excited and the fluorescence from that volume is collected as a function of time. If the number of fluorescent molecules in that volume is not changing and if there are no chemical reactions that could change the quantum yield of the fluorescence, then the average number of the emitted photon is constant. However, the instantaneous number of photon detected is not constant due to the Poisson nature of the emission/detection process. This added shot-noise is independent of time. Instead, if the number of molecules in the excitation volume is changing or the quantum yield is changing, the fluorescence intensity will change with time which is characteristic of the processes that cause the change in the fluorescence intensity. For example, if the number of molecules change due to the diffusion of a molecule out of the excitation volume, the characteristic time of this process causes characteristic frequencies to appear in the fluorescence intensity recording. Furthermore, assume that we have four molecules in the excitation volume and one leaves, the relative change in intensity will be one-fourth. However, if we have 100 molecules in the excitation volume and one leaves, the relative change will be only 1/100. Therefore the ratio of the fluctuation to the average signal is smaller the larger is the number of molecules in the volume. It can be shown that this ratio is exactly proportional to the inverse of the number of molecules in the volume of excitation (9). This relationship allows the measurements of the number of molecules in a given volume in the interior of cells (7).

A more interesting and common problem arises when molecules of different kind are simultaneously present in the same volume, either because of molecular heterogeneity or because of molecular reactions (10-12). One case of particular importance is when to macromolecules will come together to form a molecular aggregate. Let us assume that two identical proteins with one fluorescent probe each form a molecular dimer. This molecular species is different form the monomers because in carries twice the number of fluorescent moieties. When this aggregate enters the volume of excitation, it will cause a larger fluctuation of the intensity than a single monomer. Clearly, the amplitude of the fluctuation carries information on the brightness of the molecule.

On the basis of the previous discussion, the statistical analysis of fluctuations of the fluorescence signal must be done to recover the underlying molecular species and the dynamic processes that cause the change of the fluorescence intensity. It is customary to analysis the characteristic time of the fluctuation using the so called autocorrelation analysis. In this case, the autocorrelation function of the fluctuation intensity provides both the characteristic times of the system under exam and the number of fluorescent molecules in the excitation volume. In the case of identical molecules undergoing random diffusion in a Gaussian illuminated volume the characteristic autocorrelation function is given by the following expression (9):

$$G(\tau) = \frac{\gamma}{N}\left(1 + \frac{8D\tau}{w_r^2}\right)^{-1}\left(1 + \frac{8D\tau}{w_a^2}\right)^{-1/2}$$

where D is the diffusion constant, w_r and w_a are the beam waist in the radial and in the axial directions, respectively, N is the number of molecule in the volume of observation, γ a numerical factor that accounts for the non uniform illumination of the volume and τ the delay time. Other formulas have been derived for the Gaussian-Lorentzian illumination profile (7) and for molecules diffusing on a membrane (9).

The expression for the statistics of the amplitude fluctuations is generally given under the form of the histogram of the photon counts for a given sampling time Δt. This is known as the photon counting histogram (PCH) distribution. The analytical expression for the PCH distribution for a single molecular species of a given brightness has been derived for the 3D-Gaussian illumination profile (11) and is reported below.

$$p_{3DG}(k; V_0, \varepsilon) = \frac{1}{V_0}\frac{\pi \omega_o^2 z_0}{2k!}\int_0^\infty \gamma\left(k, \varepsilon e^{-4x^2}\right) dx \quad, \text{ for } k > 0$$

In this expression, V_o is the volume of illumination, ε is the brightness of the molecule and k is the number of photons in a give time interval. The integral, which contains the incomplete gamma function γ, can be numerically evaluated. Similar expressions have been derived for other shapes of the illumination volume (11). Before the development of the PCH, Qian and Elson (12) studied the effect of the intensity distribution using the

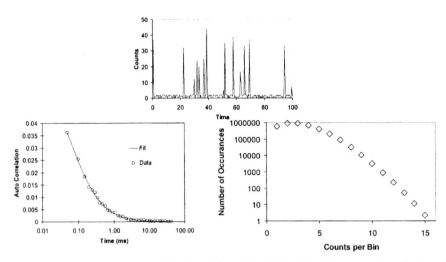

Figure 1. Upper panel: counts as a function of time. This is the original time trace data. Left panel: autocorrelation function of the time trace data from upper panel. Right panel: photon counting histogram of the time trace data in first panel.

so-called moment analysis distribution method.

A typical example of the time sequence of the fluorescence fluctuations and the calculation of the autocorrelation function and of the PCH distribution is shown in figure 1. The autocorrelation function provides the diffusion constant and the number of molecules N in the excitation volume. The PCH distribution provides the molecular brightness and the number of molecules also. In case of molecular aggregation, the autocorrelation function and the PCH distribution is fitted to a model for two or more species of different molecular brightness and of different diffusion constant (11). The particle size affects the autocorrelation function by shifting the autocorrelation curve to longer delay times for larger particle sizes. Figure 2 shows this effect for typical values of the diffusion constant of the GFP molecule (Green Fluorescent Protein) and fluorescein. The curve at smaller delay times is typical of a small molecule such as fluorescein in water (Diffusion constant of 300 $\mu m^2/s$). The next curve is typical of GFP in solution (Diffusion constant of 90 $\mu m^2/s$) and the curve on the lower panel is for a putative dimer of GFP (Diffusion constant of 70 $\mu m^2/s$). The difference between the monomer and the dimer is very small and difficult to detect in the presence of other factors, such as the changes in viscosity in the interior of cells.

For the PCH distribution, the effect of increasing the molecular brightness is that of shifting the curve to larger count number (figure 2, right panel). It is interesting to consider what happens if we mix two fluorophores of different brightness in the same sample. Theory predicts that we should obtain the convolution of the individual histograms rather than the sum. This is clearly shown in figure 3.

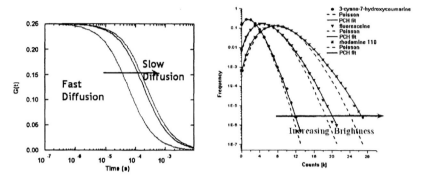

Figure 2. Left panel: Changes in position of the autocorrelation function as the value of the diffusion constant is decreased. The curve at the left is for fluorescein, the next curve is for GFP and the last curve is for a dimer of GFP molecules. Right panel: change in shape of the photon counting histogram as the brightness of the molecule is increased

The sum of the two histograms would have given two distinct distributions, but the experiments show that there is only one broad distribution which is the convolution of one PCH distribution with that of the other species. In this experiment, the molecules for sample 1 have a brightness of 5,600 counts/second per molecule (cpsm) and the concentration is about 1.08 molecules in the excitation volume as an average. For sample 2, the brightness is about 12,000 cpsm and the concentration is about 0.96 molecules in the excitation volume as an average. An equal volume of the two samples ware mixed together for the mixture sample.

The number occupancy fluctuations for each species in the mixture becomes a convolution of the individual specie histograms. The resulting histogram is then broader than expected for a single species.

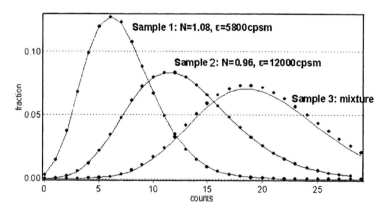

Figure 3. The convolution effect of adding two molecular species with different brightness. The mixture is not the sum of the two photon counting histograms but rather the convolution of the two distributions. N is the number of molecules in the excitation volume and e is the brightness of the molecules in units of counts/s per molecule.

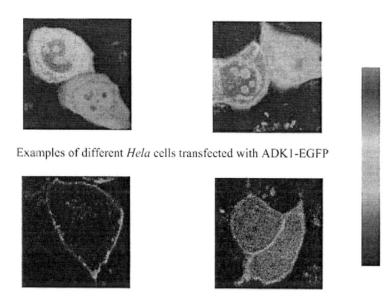

Examples of different *Hela* cells transfected with ADK1-EGFP

Examples of different *Hela* cells transfected with ADKβ-EGFP

Figure 4. Cells expressing the ADK1-EGFP chimera protein (two upper panels) and cells expressing the ADK1β-EGFP chimera protein (lower panels).

Fluctuations in cells: Protein-membrane interactions

In this paragraph, we illustrate the application of the FCS technique for the determination of the diffusion constant of EGFP (Enhanced GFP) and EGFP constructs in cells. Hela cells were transfected to produce the EGFP protein and a construct of EGFP with two different variants of the adenylate kinase protein as described in Ruan et al. (13). As the images figure 4 show, the EGFP-ADK1 protein is distributed everywhere in the cell, while the construct of adenylate kinase EGFP-ADK1β is preferentially located on the membrane of the cells.

In figure 5, the autocorrelation curve to the left is for the EGFP protein in solution. The diffusion constant corresponding to this curve is 90 $\mu m^2/s$. This value corresponds exactly to what should be expected given the molecular weight of the protein and the viscosity and temperature of the experiment (14). The next curve toward the right at longer delay times corresponds to the same protein but in the cytoplasm of the Hela cells. The value of the diffusion constant is now strongly decreased, presumably due to the larger viscosity of the cytoplasm. The next two curves, almost superimposed, correspond to the two constructs of EGFP with ADK1 and ADK1β. The two proteins are identical except for the addition of a 18-aminoacid peptide for the ADK1β protein. These proteins diffuse in the cytoplasm with an apparent diffusion constant of about 13 $\mu m^2/s$.

If we focus the laser beam on the cell membrane, the autocorrelation function shape changes dramatically. The form of the autocorrelation function is typical of that of two diffusing components as shown in figure 6, above. If the laser beam is focused in different points in the cytoplasm of the same cell, we obtain a series of values of the diffusion constant which are different in different position in the cell (figure 7). This study demonstrates that the interior of the cell is highly heterogeneous from the point of view of the diffusion of protein molecules. The heterogeneity of the diffusion could be due to interactions of the protein with other cellular components which results in slowing the motion of the protein.

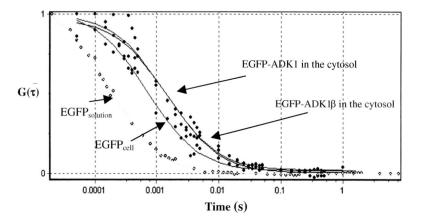

Figure 5. Autocorrelation curves for EGFP in solution and in the cytoplasm of Hela cells (two left curves) and for the chimera protein EGFP-ADK1 and EGFP-ADK1β (two right curves).

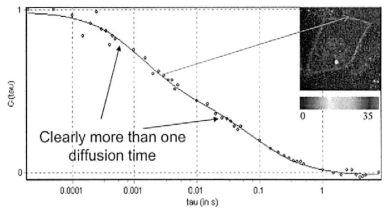

Figure 6. Autocorrelation curve for EGFP-ADK1β when excitation volume is placed on the plasma membrane. Two diffusion constants are clearly distinguished with values of 13 and 0.18 m^2/s.

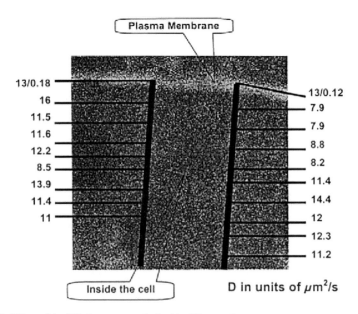

Figure 7. Values of the diffusion constants obtained in different points in the cell. The values in units of m²/s are given. When the measurement is done in the region of the cytoplasmatic membrane, two characteristic values for the diffusion constant are obtained and reported.

Cross-correlation methods

Until now, we have considered fluorescent molecules of the same color and same intensity. If there are two molecular species that differ by color and/or intensity, it is possible to detect the emission using two independent detectors that are sensitive to the two different colors. Then, the statistical analysis can be done taking into account the correlation of the fluctuations in the two detector channels. This technique is called cross-correlation (15). Fluctuation cross-correlation can provide information that is not attainable using a single detection channel. The statistical analysis is performed along similar lines as already described for one channel detection. In the dual channel experiment, we calculate the cross-correlation between the signals from the two channels and we can also construct two-dimensional photon counting histograms. In this presentation, we are only discussing the cross-correlation function and one application for the detection of internal protein dynamics. The following expression mathematically defines the cross-correlation function between two signals F1 and F2 which vary as a function of time.

$$G_{ij}(\tau) = \frac{\langle dF_i(t) \cdot dF_j(t+\tau) \rangle}{\langle F_i(t) \rangle \cdot \langle F_j(t) \rangle}$$

Let us consider the situation shown in figure 8, in which we have two kinds of

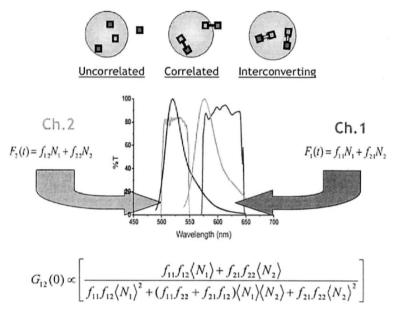

Figure 8. Upper panel: schematic representation of different possibilities of molecular species which are uncorrelated, correlated and inter-converting. Middle panel: Spectral cross-talk for the specific filters used in the experiments described in this contribution. Lower panel: The cross-correlation G(0) term for the case in which there is cross-talk between the two channels.

molecular species. If the molecular species are uncorrelated, the common (to the two channels) fluctuation on the average will reduce to zero. Therefore, if the cross-correlation is zero, we can conclude that the molecular species are independent.

If the two molecular species are connected one to the other, every time the intensity in one channel changes, it will also change in the other channel. The two signals will be fully correlated. There is a third important situation in which one molecular species changes into the other. This physical process will give rise to an anti-correlation between the signal from the two channels since when one signal increases the other will decrease and *vice versa*.

The cross-correlation at delay time zero $G_{12}(0)$ depends only on the relative contributions of the two species on the two channels. This quantity depends on the relative brightness and number of molecules present in the volume of excitation as shown in the formula in figure 8, above.

This formula is important because it shows that even if there is no physical correlation between two molecules, they could appear to be correlated because of the non-perfect separation of the fluorescence of one molecule from the other. In most practical situations it is impossible to completely separate the emission of one molecule from the other since the red tail of the emission of the bluer molecule inevitably superimpose with the emission of the redder molecule. However this effect can be accounted for by measuring the spectral cross-talk between the two channels.

Cross-correlation and Molecular Dynamics

To demonstrate that the cross-correlation measurement can provide information on internal protein dynamics we studied the cameleon protein. The cameleon protein that was used is a fusion protein consisting of calmodulin and the calmodulin binding peptide M13 sandwiched between cyan fluorescent protein (CFP) and yellow fluorescent protein (YFP). This protein was first constructed by Miyawaki et al. (16) as schematically shown in figure 9. In the presence of Ca^{2+} the CaM bends in such a way as to position the GFP constructs closer together allowing for an increase in the FRET (Förster Resonance Energy Transfer) efficiency (figure 9, right panel). FRET occurs between the CFP and the YFP when their distance became smaller than the characteristic FRET distance. Since the CaM protein is more compact when calcium is bound, this state has large FRET efficiency. To switch the cameleon into the low FRET efficiency conformation 30 mM EDTA was added (figure 9, right panel) to remove the calcium.

The important question we want to answer in this experiment is whether or not the FRET efficiency changes with time due to internal protein dynamics. The FRET efficiency is a function of the relative distance between the donor and the acceptor as well as of their relative orientation. If the FRET efficiency changes, then the spectrum will shift from blue to yellow and *vice versa*. We predict that the typical signature of the anti-correlation due to species inter-conversion between the two channels should be visible if there are dynamics changes in the FRET efficiency during the time the molecules transit across the excitation beam. This effect is shown schematically in figure 10A.

The donor autocorrelation exhibits the expected extra relaxation under the calcium saturated conditions (figure 10B, bottom). After the addition of calcium, the cross-correlation curve contains the anticipated anti-correlation (figure 10B, bottom) although it is less pronounced than in figure 10B, top, where the expected curve for this process has been simulated. This is likely a result of a large fraction of non-switching particles. Nonetheless, the anti-correlation effect is clearly visible. After the addition of dynamics yields a very small contribution of the relaxation part, indicating that

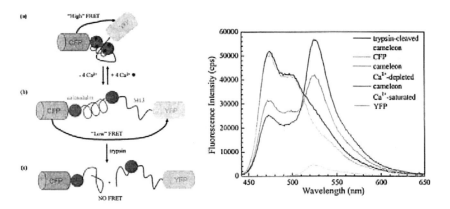

Figure 9. Left panel: schematic of the close or high FRET state in the presence of calcium and of the extended low FRET state in the absence of calcium. Right: different spectra obtained from the trypsin cleaved cameleon protein and the limiting spectra at high and low calcium concentration.

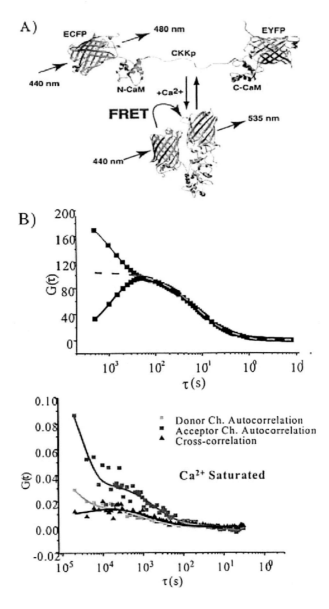

Figure 10. Top panel (A) shows the high FRET close conformation and the low FRET open conformation. Panel B, middle graph, shows the expected effect on the autocorrelation and cross correlation curves if the opening and closing reaction occurs with a characteristic relaxation time of 20 microseconds. The bottom graph shows experimental data for the auto and cross-correlation. The cross-correlation shows the characteristic anti-correlated behavior. The characteristic inter-conversion time was 20 microseconds.

a pure diffusion fit is adequate to describe the data. When calcium is added, this relaxation component increases. The curve at low calcium concentration is also much noisier than the one obtained under the calcium saturated condition. The acceptor autocorrelation curve exhibits the extra relaxation for the calcium saturated condition and cannot be fit even to pure diffusion when the EDTA is added (figure 10B, top). This is because after the EDTA is added the FRET efficiency drops to zero percent and since there is no appreciable direct excitation of YFP, the entire signal is lost.

This effect due to the protein dynamics could be exploited to determine the concentration of calcium in the cell. Of course, this protein has been engineered to have spectral sensitivity to calcium. However, for the study of the interior of cells, this relaxation behavior could also be sensitive to calcium concentration. The relaxation behavior is also present in the autocorrelation curve, showing that single channel measurements could be enough for the determination of calcium concentration.

Conclusions

We have shown that the statistical analysis of fluctuations of the fluorescence signal is a powerful tool for the study of chemical reactions both in solutions and in the interior of cells. The crucial requirement for the success of the fluctuation experiments is a means to confine the excitation of the fluorescence to a small volume, on the order of a fraction of a femtoliter. This confinement of the excitation can be achieved using confocal pinholes or two-photon excitation methods. Two-photon excitation is particularly benign to cells since out of focus photo bleaching is avoided. For solution studies, this advantage is less important. The study of the reactions in the cell interior has become a very active field of research. The dynamics of the cell interior is fundamental for life. Optical methods, in particular the far-field methods described in this presentation, provide a non-invasive way to observe the cell in action with minimal perturbation. The cross-correlation experiments described in this work for the measurement of internal protein fluctuations are also unique because they provide a relatively simple method to access the time range in the microsecond time scale which is very difficult to obtain using rapid mixing. Cross-correlation provides information on internal dynamics in the microsecond to millisecond range and it is applicable to the interior of cells. Two-photon excitation simplifies problems associated with color aberration since a single excitation wavelength can be used to excite different fluorophores.

Acknowledgements

This work was performed using the instrumentation of the Laboratory for Fluorescence Dynamics, a national research resource funded by the National Institutes of Health, NCRR, grant PHS 5 P41 RR03155 and the University of Illinois.

References

1. Elson, E.L. and Magde, D. 1974. Fluorescence correlation spectroscopy. I. *Conceptual Basis and Theory.* *Biopolymers.* 13:1-27.

2. Magde, D. 1976. Chemical kinetics and fluorescence correlation spectroscopy. *Q. Rev. Biophys.* **9**:35-47.
3. Magde, D., Elson E. and Webb, W.W. 1972. Thermodynamic fluctuations in a reacting system: Measurement by fluorescence correlation spectroscopy. *Phys. Rev. Lett.* **29**:705-708.
4. Magde, D., Elson, E.L. and Webb, W.W. 1974. Fluorescence correlation spectroscopy. II. An experimental realization. *Biopolymers.* **13**:29-61.
5. Magde, D., Webb, W.W. and Elson, E.L. 1978. Fluorescence correlation spectroscopy. III. Uniform translation and laminar flow. *Biopolymers.* **17**:361-376.
6. Denk, W., Strickler, J.H. and Webb, W.W. 1990. Two-photon laser scanning fluorescence microscopy. *Science.* **248**:73-76.
7. Berland, K.M. 1995. *Two-photon fluctuation correlation spectroscopy: method and applications to protein aggregation and intracellular diffusion.* University of Illinois at Urbana-Champaign, Urbana.
8. Berland, K.M., So, P.T.C., and Gratton, E. 1995. Two-photon fluorescence correlation spectroscopy: Method and application to the intracellular environment. *Biophys. J* **68**:694-701.
9. Thompson, N.L. 1991. Fluorescence correlation spectroscopy. In *Topics in Fluorescence Spectroscopy.* Lakowicz, J.R. (ed.), Plenum, NY 337-378.
10. Palmer, A.G. and Thompson, N.L. 1987. Molecular aggregation characterized by high order autocorrelation in fluorescence correlation spectroscopy. *Biophys J* **52**:257-270.
11. Chen, Y., Muller, J.D., So, P.T. and Gratton, E. 1999. The photon counting histogram in fluorescence fluctuation spectroscopy. *Biophys J.* **77**(1):553-567.
12. Qian, H. and Elson, E.L. 1990. Distribution of molecular aggregation by analysis of fluctuation moments. *Proc Natl. Acad. Sci. U S A.* **87**:5479-5483.
13. Ruan, Q., Chen, Y., Gratton, E., Glaser, M. and Mantulin, W.W. 2002. Cellular characterization of adenylate kinase and its isoform: two-photon excitation fluorescence imaging and fluorescence correlation spectroscopy. *Biophys J* **83**(6):3177-3187.
14. Chen, Y., Muller, J.D., Ruan, Q., Gratton, E. 2002. Molecular brightness characterization of EGFP *in vivo* by fluorescence fluctuation spectroscopy. *Biophys J* **82**(1):133-44.
15. Schwille P, Meyer-Almes, F. J. and Rigler, R. 1997. Dual-color fluorescence cross-correlation spectroscopy for multicomponent diffusional analysis in solution. *Biophys J* **72**: 1878-1886
16. Miyawaki, A., Llopis, J., Heim, R., McCaffery, J. M., Adams, J. A., Ikura, M. and Tsien, R. Y. 1997. Fluorescent indicators for Ca2+ based on green fluorescent proteins and calmodulin. *Nature (London)* **388**, 882-887.

Chapter 2

DYNAMICS OF THE CELL MEMBRANE OBSERVED UNDER THE EVANESCENT WAVE MICROSCOPE AND THE CONFOCAL MICROSCOPE

Susumu Terakawa[1], Takashi Sakurai[1], Takashi Tsuboi[1], Yoshihiko Wakazono[1], Jun-Ping Zhou[1], and Seiji Yamamoto[1]

1. INTRODUCTION

The objective lens of a microscope plays an essential role for high resolution imaging. Especially, the numerical aperture (NA) of the objective lens is the determining factor for gaining the resolving power. In 1994, a novel objective lens that had a numerical aperture of 1.65 was developed by Susumu Terakawa (the author) and Katsuyuki Abe (Olympus Optical Co.). Use of this ultra high NA lens opened a new field of light microscopy beneficial for observations of living cells at a molecular level and for studies of the cell membranes in their dynamic aspects.

2. ULTRA HIGH NA OBJECTIVE LENS

Use of the ultra high NA lens (Figure 1) is quite advantageous for observation of fluorescent objects of biological specimens under the illumination of evanescent light[1]. The ultra high NA lens has a large extra portion that accepts a laser light introduced from the back side of the lens, and projects it to the focal point at an angle of incidence larger than the critical angle (Figure 2). Such an extra portion lies only in the marginal zone between the line of critical angle and the edge of the lens.

The ultra high NA (NA=1.60) lens was once manufactured more than 100 years ago at Carl Zeiss, Jena. However, in fact, it disappeared from the market completely after Ernst Abbe pointed out that the extra large portion of the lens was useless for collecting the light from the objects immersed in the water.

---

[1] Photon Medical Research Center, Hamamatsu University School of Medicine, Hamamatsu, 431-3192 Japan
Email: terakawa@hama-med.ac.jp

Figure 1. Objective lens of 1.65 NA (Olympus) developed by Terakawa and Abe in 1994.

Since then, no such lens was commercially available, and thus the ultra high NA lens has been forgotten for a long time. Aiming at an improvement of the resolution and the maximum brightness, we newly developed an objective lens that has a numerical aperture of 1.65[1,2].

The lens has a high performance in three respects. 1) It is the brightest lens ever made in the history of optics. 2) The resolution reaches the highest level among all optics: the smallest resolvable distance is smaller than 150 nm. 3) It creates thinnest evanescent field (< 50 nm) easily when used for a through-the-lens (TTL)[3] evanescent wave illumination. All of these properties combined together, it is the most advantageous lens for the microscopy with an evanescent wave illumination or total internal reflection fluorescence microscopy (TIRFM)[1]. A disadvantage is that one has to use the lens in combination with specially selected oil and a coverslip of an optimized refractive index of 1.78.

This 1.65NA lens collects a light with its acceptance angle larger than the so called critical angle. No possible optical path exists for the light coming from the water phase in any direction illuminating the focal point O (in Figure 2). However, if the light source is smaller than the wave length of light, and if it is located exactly on the interface between the water and the coverglass i.e., the first optical element collecting the light from the source, then the light from the source radiates unevenly with a peak radiation in the direction of the critical angle, forming a distribution pattern as shown in Figure 3. A significant intensity of light is to be detected in the direction beyond the critical angle[4]. This is easily understood because the evanescent light generated in an area of 50 – 100 nm distant from the interface can travel into the glass phase in the direction beyond the critical angle. The light generated at a source located in an even shorter distance can follow a similar optical path.

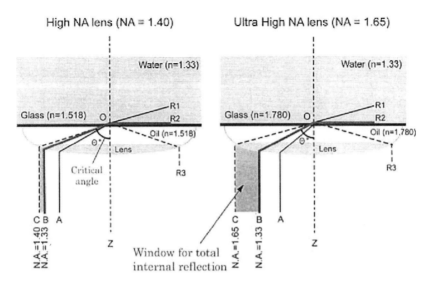

Figure 2. Ultra high NA lens focusing on a small object on the interface between the water (top half) and the glass (bottom half). A 1.40 NA lens (left) and a 1.65 NA lens (right) are compared schematically. An ordinary optical path that starts from a point (R1) in the water phase forms a line indicated by A. An optical path with the critical angle at the interface is indicated by B, and with a largest possible angle by C.

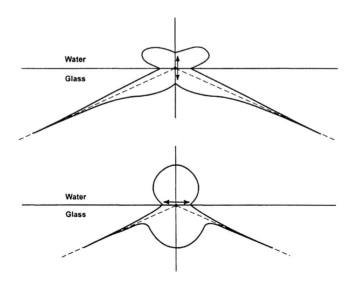

Figure 3. Radiation of light from a point source on the interface between water and glass. The dipole (arrows) lies perpendicular (top) and parallel (bottom) to the interface. The dashed line indicates the critical angle. (Hellen and Axelrod, 1987)[4].

3. OBSERVATIONS UNDER EVANESCENT WAVE ILLUMINATION

The evanescent field formed by the TTL method is useful for visualizing the molecular image of fluorescent dyes as well as the molecular dynamics of the cell membranes.

3.1. Calcium Indicator Dye

Figure 4 shows images of a calcium sensitive dye, Rhod-2, stuck on the surface of glass. When the Ca^{2+} was chelated from the medium by addition of EGTA, no significant fluorescence was detected. However, when the Ca^{2+} concentration was raised to a level as high as 130 nM, discrete spots of fluorescence were clearly observed. A calcium dependent increase in fluorescence can be observed at the molecular level. Loading this dye into neurons made it possible to monitor Ca^{2+} signals near the plasma membrane or in the vicinity of the Ca^{2+} channels. Imaging the Ca^{2+} transients exclusively near the plasma membrane with a CCD camera of the fast rate (5 ms) was sensitive enough to detect some clustered responses.

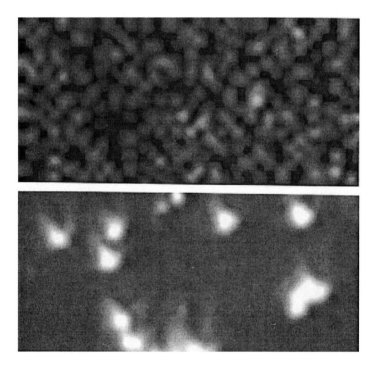

Figure 4. Fluorescence image of Rhod-2 dye molecules captured under the evanescent wave microscope equipped with a 1.65 NA lens. When [Ca^{2+}] is low, almost no fluorescence was observed (top). When [Ca^{2+}] is 130 nM, single molecule images appeared brightly (bottom).

3.2. Ion Channel

Labeling of ion channels with a fluorescent dye can be selectively performed by modification of the RNA code and by injection of modified RNA into a model cell like the *Xenpus* oocyte. The *Shaker* K channel was expressed in the oocyte after code modification of S351C. After the expression of proteins, the cystein residue was labeled with tetramethyl rhodamine maleimide. The cells placed under evanescent wave illumination showed many small fluorescent spots. Depolarization of the plasma membrane by 120 mV by the double electrode voltage clamp technique induced a large change in fluorescence intensity of these fluorescent spots (Figure 5). Many of them showed such a voltage-dependent fluorescence change in addition to the quantized bleaching of fluorescence, indicating that the spot was a single molecule of the K channel or at least the subunit of the channel[5]. These single channel images were rather static, and no sign of the fast mobility in lateral direction was observed.

3.3. Protein Kinase-C

The evanescent field decays exponentially with the distance from the surface of glass, with decay constant being the function of the ratio of refractive index of water to that of glass and also the function of the angle of incidence and the wavelength of the light. The fluorescent object entering the field from the outside will increase the fluorescence intensity dramatically. This property can be exploited for visualization of activated form of some enzymes. For example, protein kinase C (PK-C) is translocated to the plasma membrane from the cytoplasm upon activation by phorbol ester. When a HeLa cell expressing GFP fused with PK-C was stimulated with phorbol ester, a dramatic increase in fluorescence was observed first in a short time reversibly at 2 – 3 min, and then delayed and persistent manner after 10 min (Figure 6).

A similar response appeared clearly but in a much smaller degree under a confocal microscope.

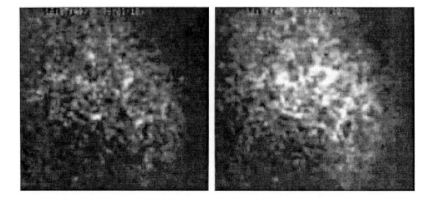

Figure 5. Tetramethyl rhodamine-labeled *Shaker* K-channel expressed in a *Xenopus* oocyte. The membrane voltage was varied from -80 mV (left) to +40 mV (right) by the voltage clamp method.

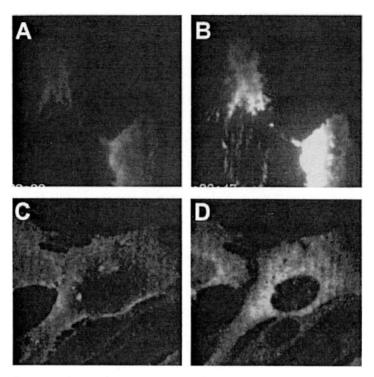

Figure 6. Fluorescence images of cells expressing PK-C/GFP observed under evanescence mode (top) and confocal mode (bottom), before (left) and 9 min after (right) stimulation with TPA.

3.4. Dynamin

The dynamin activity can also be observed in a similar manner when the protein is fused with GFP. Stimulation of a cell (PC12) with an electrical pulse induced a translocation of the protein, i.e., many clusters of the dynamin-GFP fusion protein appeared on the plasma membrane in 30 – 90 s after stimulation. These fluorescent clusters showed rather rapid Brownian movement under the plasma membrane. Some of them were lined up and moved a few micrometer as if they swept up scattered debris[6]. In rare cases, hairy clusters were assembled into a circular pattern which enlarges itself in a minute or so as shown in Figure 7. These dynamic images provide evidence that the dynamin interacts continuously from one pit (or invagination) to the other on the plasma membrane. Probably dynamin pinches off the pit membranes at the neck without its detachment from the plasma membrane. We hypothesized that the dynamin action takes place just like a sweeper cleaning on the floor in a scanning manner (sweeping model). This is in some contrast to the conventional view that dynamin binds to the neck region and detaches the plasma membrane together with a pit membrane to form an endocytic vesicle.

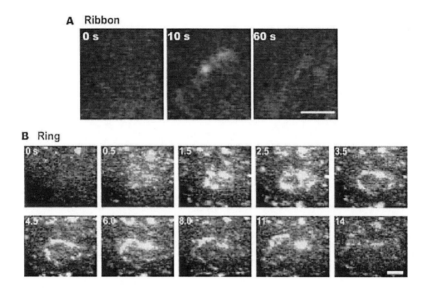

Figure 7. Super clusters of dynamin visualized in PC12 cells expressing the dynamin-GFP fusion protein[6]. The cell was stimulated with an electrical pulse of 1 ms in duration, and then video sequences were captured. The numbers indicate the time after the stimulation in second. Bars, 1 μm.

3.5. IP$_3$

Some of the intracellular messengers involve translocation of signal molecules from the plasma membrane to the cytoplasm. As a result of enzymatic reactions, activation of G-protein and phospholipase C for example, IP$_3$ is produced on the plasma membrane and is released to further activate Ca^{2+} channels located intracellularly on the membranous structures. This process can be monitored by a GFP fusion protein capable of binding the so called PH domain. When phosopholipids are in their original form, the protein binds to the PH domain on the phospholipids. After cleavage of the phospholipids, the protein loses its binding site on the plasma membrane. Under the evanescent wave illumination, a large loss of membrane fluorescence was observed in HeLa cells expressing the PH-domain binding protein fused to GFP when the cell was stimulated with histamine. Sometimes, waves of darkness propagated across the bottom part of the cells, indicating a repeated cycle of IP$_3$ signals in a few minutes.

3.6. Exocytosis

Exocytosis is a very dynamic process taking place on the cell membrane. When contents or the vesicle membrane is stained with fluorescent dyes, the process can be observed at a high spatio-temporal resolution. Adrenergic vesicles are easily stained with acridine orange, as it mimics the adrenaline or noradrenaline in the uptake process on the vesicle membrane. Expression of a GFP fusion protein either as a content of secretory

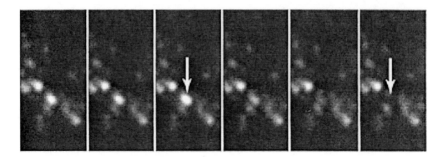

Figure 8. Exocytosis of an acridine orange-loaded secretory granule in a MIN-6 cell. The cell was stimulated by irradiation of an UV laser flash after loading with caged Ca^{2+}. A sequence of video images were recorded at a frame rate of 33 ms. The granule (left arrow) transiently increased the fluorescence and then rapidly disappeared (right arrow). Almost all granules shown in this figure responded similarly in a few second.

vesicles or as an intrinsic protein on the limiting membrane is an obvious alternative to visualize the exocytotic response under the evanescent light[7].

The exocytotic response of fluorescent vesicles appeared in a two different manners[8]. The vesicles rapidly disappeared one by one when the cell was stimulated with agonists or by membrane depolarization. There was no optical sign before the disappearance. The fluorescence intensity decreased to zero level in 30 – 100 ms in the chromaffin cells and MIN-6 cells. In the case of response of transmitter release from the nerve terminals, the rate of decrease reached 6 ms which was the smallest time resolved by our custom made CCD camera. Another form of response was a transient flash response immediately followed by disappearance as mentioned above (Figure 8). The flash was an increase in fluorescence intensity as well as in fluorescent area. In the evanescent field, the intensity of excitation light decreases with the distance from the surface. Because of this property, fluorescent objects increase in intensity as they move closer to the glass surface. We interpret that two types of exocytotic responses found in a single cell reflect the variation in the velocity of releasing process. When the velocity is faster than the diffusion rate, the concentration of substances released would increase near the glass surface transiently before they diffuse away. This is why the flash response appears before the disappearance of vesicles. The release of substance from a vesicle by diffusion process only would appear as simple disappearance of the vesicle. The extra factor that facilitates release is a volume flow of water due to secretion from the vesicles[8].

4. DISCUSSION

The evanescent wave microscope can be applied to molecular visualization of membrane proteins such as ion channels, receptors, and transporters. It is also useful for dynamic analyses of the transmembrane signaling, release of signal molecules, exocytosis, and endocytosis. The evanescent wave illumination provides very localized uncaging of Ca^{2+} and other caged compound. The resulting distribution of Ca^{2+} may be very physiological similar to that induced by opening of Ca^{2+} channel upon the membrane depolarization. Ca^{2+} transients near the cell membrane is also selectively monitored by this evanescent wave illumination. As the evanescence microscope can be assembled easily even by beginners, this field of research is very promising in future.

5. ACKNOWLEDGEMENTS

We thank Miss. A. Takase, Mrs. Y. Gotoh, and Mrs. N. Nakamura for their technical assistances. The present study was supported in part by the Grant-in-Aid for Scientific Research B (14370010) to ST and the Grant-in-Aid for Intelligent Cluster Project to ST from the Ministry of Education, Culture, Sports, Science and Technology, Japan.

6. References

1. Kawano, Y., Abe, C., Kaneda, T., Aono, Y., Abe, K., Tamura, K., Terakawa, S., 2000, High numerical aperture objective lenses and optical system improved objective type total internal reflection fluorescence microscopy, *Proc SPIE* 4098: 142-151.
2. Terakawa, S., Tsuboi, T., Sakurai, T., Jeromin, A., Wakazono, Y., Yamamoto, S., Abe, K., 2001, Fluorescence micro-imaging of living cells and biomolecules with ultra high NA objectives, *Proc SPIE* 4597: 121-127.
3. Tokunaga, M., Kitamura, K., Saito, K., Iwane, A.H., Yanagida, T., 1997, Single molecule imaging of fluorophores and enzymatic reactions achieved by objective-type total internal reflection fluorescence microscopy, *Biochem Biophys Res Commun.* 235: 47-53.
4. Hellen, E.H., Axelrod, D., 1987, Fluorescence emission at dielectric and metal-film interfaces. *J Opt Soc Am B* 4: 337-350.
5. Sonnleitner, A., Mannuzzu, L.M., Terakawa, S., Isacoff, E.Y., 2002, Structural rearrangements in single ion channels detected optically in living cells, *Proc Nat Acad Sci USA* 99(20): 12759-64.
6. Tsuboi, T., Terakawa, S., Scalletar, B., Fantus, C., Roder, J., Jeromin, A., 2002, Sweeping model of dynamin activity, *J Biol Chem.* 277: 15957-61.
7. Tsuboi T, Zhao C, Terakawa S, Rutter GA, 2000, Simultaneous evanescent wave imaging of insulin vesicle membrane and cargo during a single exocytotic event, *Curr Biol.* 10: 1307-1310.
8. Tsuboi, T., Kikuta. T., Sakurai, T., Terakawa, S. 2002, Water secretion associated with exocytosis in endocrine cells revealed by micro forcemetry and evanescent wave microscopy. *Biophys J.* 83: 172-83.

Chapter 3

USING GFP AND FRET TECHNOLOGIES FOR STUDYING SIGNALING MECHANISMS OF APOPTOSIS IN A SINGLE LIVING CELL

Donald C. Chang[1], Liying Zhou[1], and Kathy Q. Luo[1,2]

1. INTRODUCTION

One of the most important problems in life science today is to understand the control mechanisms of cell cycle and programmed cell death (also called "apoptosis"). Programmed cell death is a controlled cell suicidal process that has great importance in maintaining the normal physiological function of a human being. It allows the biological organism to destroy damaged or unwanted cells in an orderly way. For example, apoptosis is used in the thymus to eliminate self-reactive T cells to avoid auto-immunity (Thompson, 1995). Furthermore, apoptosis plays a critical role in animal development; it is utilized to eliminate the unwanted cells so that the proper structure of functional organs can be formed in the embryo. For example, formation of digits in a hand and the proper targeting of the neural connection are known to be dependent on apoptosis. Thus, defects in apoptosis will result in major developmental abnormalities (Jacobson et al., 1997).

The study of apoptosis has gained widespread attention in recent years due to its newly discovered roles in a variety of pathological processes. For example, when DNA is damaged in a cell and cannot be repaired, the cell will enter apoptosis to avoid the formation of abnormalities in the tissue. Thus, failure of programmed cell death can cause cancer. Besides tumorgenesis, malfunction of apoptosis is also found to be associated with numerous pathological disorders, including Alzheimer's disease, auto-immune disease, AIDS, etc.

In the last few years, a large number of studies have been conducted aiming to understand the process of apoptosis on a molecular basis. The signaling pathways that direct the programmed cell death process have turned out to be very complicated. There

---

[1] Department of Biology, the Hong Kong University of Science and Technology, Clear Water Bay, Kowloon, Hong Kong, China. bochang@ust.hk; [2] Department of Chemical Engineering, the Hong Kong University of Science and Technology, Clear Water Bay, Kowloon, Hong Kong, China.

are many external signals that can trigger the initiation of apoptosis, including UV-irradiation, activation of the "death domain" via the TNF (Tumor Necrosis Factor) receptor, treatment of hormone (e.g. glucocorticoid) and chemotherapy drugs (e.g. camptothecin) (Martin and Cotter, 1991; Nagata, 1997) (see Figure 1). As for internal signals, we know that apoptosis is the outcome of a programmed cascade of intracellular events, which are centered on the activation of a class of cysteine proteases called "caspases" (Cohen, 1997). Some of these caspases (such as caspase-8 and caspase-9) are "initiators" of the apoptotic process, while others (such as caspase-3) are "executioners". Besides caspases, a number of gene products are also known to be key players in processing apoptosis. These include the Bcl-2/Bax family, p53, IAP (Inhibitor of Apoptosis Protein) and Smac/Diablo (Du et al., 2000; Oren and Rotter, 1999; Reed et al., 1998). When these genes are over-expressed (or mutated), some of them can cause cells to undergo apoptosis, while others can prevent cells from entering programmed cell death. Some of the organelles are also known to play an important role in the progression of apoptosis (Desagher and Martinou, 2000). Among them, mitochondria plays the most important role in inducing apoptosis, since some of the substances discharged from mitochondria (including cytochrome c, Smac, AIF and endo G) are key activators of the downstream apoptotic pathways (Desagher and Martinou, 2000; Wang, 2001). A summary of the major signaling pathway of apoptosis is shown in Figure 1. At present, the detailed molecular mechanisms of apoptosis regulated by various external and internal signals are still under active investigation.

In recent years, our laboratory has been focusing on developing innovative optical techniques for the study of these molecular signaling systems in a single living cell. These techniques include labeling specific proteins with GFP (Green Fluorescent Protein)

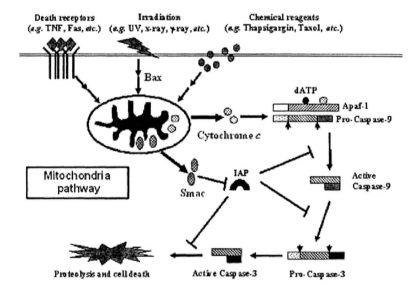

Figure 1. Schematic model of the mitochondria-dependent signaling pathway in apoptosis.

using gene-fusion, and measurements of protein-protein interaction using the FRET (Fluorescence Resonance Energy Transfer) method. Our approach has certain advantages. First, since these optical methods are non-invasive, we can preserve the organization and structure of the cell during the biological study. Thus, the information obtained is highly reliable. Second, by conducting the study in a single cell, we can avoid the problem of cell synchrony, which would be difficult to achieve in a cell population study. Third, in the single cell study, we can correlate the temporal- and spatial-dependent changes of a specific signaling molecule with a particular cellular event. Finally, using the FRET technique, we can measure the dynamics of a specific protein-protein interaction or the activation of a given enzyme within a single cell. Such measurements are not possible using conventional biochemical methods.

In this chapter, we will give several examples to demonstrate how we use these innovative optical technologies for studying signaling mechanisms in programmed cell death. The first example is to investigate the temporal relationship between Bax translocation and the release of cytochrome c during UV-induced apoptosis. Mitochondria were known to play a central role in the apoptotic process. Specifically, releases of cytochrome c and Smac from mitochondria are the key steps that trigger the activation of caspase-9 and caspase-3 during cell death. At present, it is known that the release of mitochondrial proteins is regulated by the Bcl-2/Bax family proteins. The mechanism, however, is not yet clear. There is also a question on whether different types of apoptotic proteins were released from mitochondria through the same pathway. In this study, we have gained important insight in these problems by examining the dynamic re-distribution of GFP-labeled Bax during apoptosis using living cell-imaging techniques.

In the second example, we will discuss the development of novel molecular bio-sensors based on the FRET method for detecting activation of specific enzymes within a single living cell. These sensors were constructed by fusing a CFP (Cyan Fluorescent Protein) and a YFP (Yellow Fluorescent Protein) with a linker containing a caspase-specific cleavage site. The change of the FRET ratio upon cleavage was larger than 4 fold. Using these sensors, we were able to measure the dynamics of activation of caspase-3 and caspase-8 in a single living cell. These sensors not only enable us to study the molecular mechanisms of cell regulation in an *in vivo* condition, they also provide a powerful tool for rapid-screening of new drugs. Particularly, these FRET-based bio-sensors can be developed into a useful platform technology for discovering the active components contained in various traditional Chinese medicine (TCM).

2. USING THE GFP-GENE FUSION TECHNIQUE TO STUDY THE DYNAMIC REDISTRIBUTION OF SIGNALING PROTEINS IN A SINGLE LIVING CELL

2.1. General Properties of GFP

Green fluorescent protein (GFP) is a naturally fluorescent protein (MW 27 kDa) isolated from the jellyfish *Aequorea victoria* (Shimomura et al., 1962). In recent years, GFP has become a widely used fluorescent marker for research in molecular biology, cell biology and medicine. One major advantage of GFP is that it can become fluorescent without requiring any exogenous substrates or enzymes. This is because the chromophore of GFP is formed by an internal post-translational autocatalytic cyclization of three amino acids (Chalfie et al., 1994). This makes the GFP as an ideal fluorescent tag for living-cell imaging. Another advantage of GFP is its high stability. GFP fluorescence is very stable in the presence of denaturants or proteases, as well as over a range of pH and temperatures. GFP is also highly tolerant to many classical fixatives and fluorescence quenching agents, and thus can be easily detected in preserved tissues. In addition, the GFP chromophore is relatively photo-stable during imaging. This high stability of GFP is attributed to its unique three-dimensional structure (Figure 2A). Eleven strands of β-sheet form an antiparallel barrel with short helices forming lids on each end (Yang et al., 1996). The chromophore, formed by cyclization of 65Ser-Tyr-Gly67, is inside the cylinder. The tightly constructed β barrel thus can protect the chromophore from the surrounding environment and make the fluorescent protein highly stable. Another advantage of the highly compact structure of the GFP molecule is that, it usually does not interfere with the folding of the target protein. This makes the tagged protein to have a good chance to retain its original biochemical properties. Using random and site-directed mutagenesis, many GFP mutants that have different excitation and emission spectra have been produced in the last five years (see Figure 2B). Some of the mutants were also optimized for brighter emission, faster chromophore maturation and more resistance to photobleaching. These useful GFP mutants include: EGFP (Enhanced GFP, green), EYFP (yellow), ECFP (cyan), and EBFP (blue). In addition to GFP and its variants, a brilliantly red fluorescent protein called "DsRed" has recently been cloned from Discosoma coral. By combining the applications of DsRed and GFP variants, one can now monitor the movement of multiple tagged-proteins in a cell using multi-color imaging.

2.2. Applications of the GFP Technology for Biophotonic Studies of Programmed Cell Death

In recent years, the GFP technology has been used in a wide variety of applications, including studies on: (a) pattern of gene expression, (b) sub-cellular localization of proteins, (c) dynamic redistribution of proteins, (d) co-localization of different proteins within the same cell, and (e) protein-protein interactions through FRET analysis.

Our study showed that this GFP technology is a very powerful tool for investigating the apoptotic signaling pathways in living cells. As shown in Figure 1, the releases of two proteins, cytochrome c and Smac, from mitochondria to the cytosol have been identified to be the key steps in triggering the activation of caspase cascades in apoptosis. The released cytochrome c can induce the "apoptosis protease-activating factor" (Apaf-1) protein to form a multimeric complex to recruit and activate procaspase-9, which in turn activates procaspase-3. The released Smac, on the other hand, can neutralize a set of caspase inhibitors, known as IAP. The combination of activation of caspases and removing of their inhibitors finally causes cell death. In order to study the mechanism of cytochrome c release during apoptosis, we used a confocal microscope to monitor the dynamic re-distribution of GFP-labeled cytochrome c (Cyt c-GFP) in intact living HeLa cells following UV treatment. Figure 2C shows a sample record of this time series measurement. Before cytochrome c was released from mitochondria, it was distributed in a filamentous pattern. As cytochrome c was released, the Cyt c-GFP became uniformly distributed over the entire cytoplasm. This releasing process was found to be very fast; it usually finished in about 5 minutes.

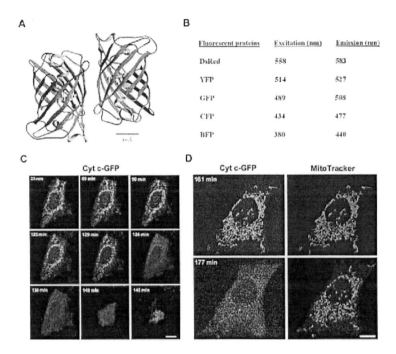

Figure 2. (A) The three-dimensional structure of GFP (Yang et al., 1996). Reprinted with permission from Nature Biotechnology. (B) Fluorescence proteins that have different excitation and emission wavelengths. (C) Sample records from a time-series confocal measurement of Cyt c-GFP distribution in a single living HeLa cell during UV-induced apoptosis. (D) Sample records from time-series confocal measurements of Cyt c-GFP distribution (left) and Mitotracker distribution (right) in a living HeLa cell during apoptosis. Elapsed time following the UV treatment is shown in each panel of C and D. Scale bar, 10 μm.

At present, the mechanism of cytochrome c release is not yet clearly understood. One leading hypothesis is that, a megachannel traversing across the outer and inner mitochondrial membranes called "PT pore" is induced to open during apoptosis (Desagher and Martinou, 2000; Gao et al., 2001). Solutes and water in the cytosol then enter mitochondrial matrix through this channel, causing the mitochondrion to swell and rupture its outer membrane. As a result, cytochrome c and other proteins in the inter-membrane space are released. This hypothesis has been widely cited and debated in the literature (Desagher and Martinou, 2000; Gao et al., 2001), but so far, there is still a lack of conclusive evidence to prove or disprove it. In order to provide a critical test of this swelling theory, we used fluorescence imaging techniques to measure the dynamic re-distribution of GFP-labeled cytochrome c in living HeLa cells, and used a red color fluorescent dye, Mitotracker, to image the morphological change of mitochondria at the same time (Figure 2D). We found that mitochondria did swell when HeLa cells were induced to enter apoptosis by UV treatment. The swelling, however, occurred only after cytochrome c was released. Results of this living cell imaging study strongly suggest that cytochrome c release in apoptosis is not caused by mitochondrial swelling.

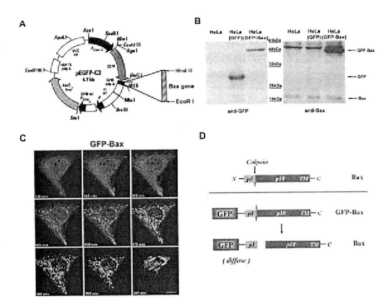

Figure 3. (A) Subcloning of Bax into a pEGFP-C3 vector. (B) The expression of GFP-Bax in HeLa cells was confirmed by Western-Blot. Cell lysate of HeLa cells transfected with either GFP or GFP-Bax gene was probed using antibodies against GFP or Bax to detect the GFP-Bax fusion protein. (C) Sample records from a time-series confocal measurement of GFP-Bax distribution in a single living HeLa cell during UV-induced apoptosis. Elapsed time following the UV treatment is shown in each panel. Scale bar, 10 μm. (D) Schematic diagram showing the experimental design for testing the model that the N-terminus cleavage by calpain is responsible for Bax activation in apoptosis.

So, we need to investigate alternative mechanisms. It has been suggested recently that formation of "channels" in the outer mitochondrial membrane by Bax-like proteins may be responsible for the release of mitochondrial proteins, such as cytochrome c and Smac (Desagher and Martinou, 2000). Several different models have been proposed: (a) Bax may form channels by itself; (b) Bax may form chimeric channels with the voltage-dependent anion channel (VDAC) protein; (c) Bax may form a "protein-lipid complex" to destabilize the mitochondrial membrane (Desagher and Martinou, 2000). In this study, we decide to test these models by examining the dynamic re-distribution of GFP-labeled Bax during UV-induced apoptosis using living-cell imaging techniques. Figure 3A shows the construction of the GFP-Bax probe. After the Bax gene was subcloned into a pEGFP-C3 vector (from Clontech), we introduced it into HeLa cells by electroporation. The expression of this GFP-Bax fusion gene was confirmed using a Western blot analysis (Figure 3B). Using this GFP-Bax as a probe, we have studied the dynamic re-distribution of Bax during UV-induced apoptosis in single living HeLa cells (Figure 3C). Bax was initially found to be localized in the cytosol in a diffused pattern. Then, during the progression of apoptosis, part of the Bax proteins was observed to translocate from the cytosol to mitochondria and began to form small punctate structures there. Soon after that, more Bax proteins were found to aggregate in mitochondria to form large clusters. About half an hour later, the cell shrank and died.

In order to answer the question whether the translocation of Bax is responsible for the release of cytochrome c or Smac, we compared the time course between Bax translocation and the release of Smac during UV-induced apoptosis by co-transfecting CFP-Bax and Smac-YFP into HeLa cells. Our results showed that the formation of CFP-Bax punctate structures in mitochondria was temporally associated with the release of Smac (data not shown). We also found that the release of cytochrome c occurred at the same time as the formation of the Bax punctate structures (data not shown). These results suggest that both cytochrome c and Smac are released from mitochondria through the Bax-formed "pores". But what is the structure of this type of "pores"? Since Smac is known to be released as a tetramer (100kD) from mitochondria into the cytosol during apoptosis, such a large protein complex is unlikely to pass through a channel formed by either Bax alone or Bax associated with VDAC. Hence, we think that the lipid-protein complex formed by Bax in the outer mitochondrial membrane is probably the most viable model for explaining the release of mitochondrial proteins during apoptosis.

2.3. Using the GFP Technology to Test a Specific Model of Bax Activation

One of the remaining questions is how Bax is activated to translocate from the cytosol to the outer mitochondrial membrane. It has been suggested that Bax may be cleaved at the N-terminus by calpain to remove a pro-domain (p3) to generate a proapoptotic Bax/p18 fragment, which translocates to mitochondria and trigger cytochrome c release (Gao and Dou, 2000). We have used the GFP-gene fusion technique to test this hypothesis. By tagging the Bax protein with a GFP at its N-terminus, we monitored the translocation of the fused protein at various times in a UV-induced apoptotic cell. If Bax was activated through cleavage by calpain, GFP should only connect to the p3 pro-domain and is separated from the active Bax (Figure 3D). This small N-terminus tagged GFP should always be in the cytosol during apoptosis, since it does not contain the transmembrane (TM) domain, which is in the C-terminus of Bax. This, however, was not what we observed. As shown in Figure 3C, GFP-Bax was found to translocate to mitochondria during UV-induced apoptosis, indicating no cleavage could happen during the activation of Bax. Thus, we must conclude that N-terminus cleavage by calpain is not responsible for Bax activation. Currently, we are continuing to use the GFP technology to investigate other mechanisms for Bax activation in apoptosis.

3. APPLICATION OF THE FRET TECHNIQUE

3.1. The Principle of FRET

Fluorescence resonance energy transfer (FRET) is a phenomenon during which the excited fluorescent energy was transferred from a donor fluorescent molecule to an acceptor fluorescent molecule located in a close proximity. The mechanism of energy transfer was first described by Forster (Forster, 1959). The principle of FRET is relatively simple: When a donor molecule is excited by absorbing a photon at certain wavelength, it can transfer its excitation energy to its neighboring acceptor molecule. As a result, the acceptor molecule is excited and emits the fluorescent light (Forster, 1959). According to Forster, the efficiency of energy transfer (E_t) is strongly dependent on the distance (r) between the donor and the acceptor:

$$E_t = \frac{R_o^6}{R_o^6 + r^6}$$

R_o depends on two major factors: (a) the degree of spectral overlap between the donor emission and the acceptor excitation, and (b) the orientation angle between the donor and the acceptor, which can be described by the orientation factor κ^2.

$$\kappa^2 = (\cos\alpha - 3\cos\beta \cos\gamma)^2 \qquad (2)$$

Here, α is the angle between the emission dipole of the donor and the absorption dipole of the acceptor, β and γ are the angles between the dipoles and the vectors joining the donor and the acceptor. Since the efficiency of energy transfer is related to the sixth power of the separation distance between the donor and the acceptor, FRET is a very sensitive tool for detecting changes of distance between two interacting molecules, which, for example, can provide important information on protein-protein interactions (Zhang et al., 2002).

3.2. Designing a Bio-sensor for Measuring Caspase-3 Activity Based on FRET

A new application of biophotonics is to insert a sensing peptide between two different colored GFPs and use the FRET method to detect conformational change of the fused protein (Heim and Tsien, 1996). We have applied this new technology to develop bio-sensors for detecting the activation of caspase-3. As caspase-3 activation is a critical step in apoptosis, assaying caspase-3 activity can be used as a tool for detecting apoptotic cell death (Cohen, 1997). At present, most of the commonly used assays are *in vitro* methods, which utilize cell extracts from a population of cells. Due to the lack of synchrony in the progression of apoptosis within a cell population, it is difficult to study the dynamics of caspase activation using the cell extract method. Recently, we have developed a special technique that uses the FRET method to detect the activation of

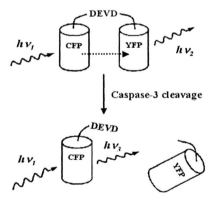

Figure 4. The principle of the FRET sensor for detecting caspase-3 activation.

caspase-3 in a single living cell (Luo et al., 2001). The basic principle of this method is illustrated in Figure 4. We constructed a molecular bio-sensor by fusing a CFP and a YFP with a specialized linker containing the caspase-3 cleavage sequence DEVD (Luo et al., 2001). Before caspase-3 was activated, the two differently colored GFPs (called donor and acceptor) are covalently linked together; energy can be transferred directly from the donor to the acceptor, so that when the donor is excited, fluorescence emitted from the acceptor can be detected. When caspase-3 is activated, it cleaves the linker and causes the donor and acceptor to separate. Thus, the FRET effect is effectively eliminated (Figure 4). This FRET probe was designated as "sensor C3".

3.3. In Vitro Characterization of Sensor C3 Using Purified Protein

In order to characterize the properties of this biosensor, we expressed the sensor C3 gene in *E. coli* and purified the polyhistidine-tagged recombinant protein using a Ni-NTA column. The purified protein was incubated with or without caspase-3 and then subjected to PAGE and Western blot analysis. The results showed that the sensor C3 protein was completely cleaved by caspase-3 into two protein bands. When a caspase-3 inhibitor (Ac-DEVD-CHO) was added, sensor C3 could no longer be cleaved, indicating sensor C3 can be cleaved specifically by caspase-3 (Luo et al., 2001).

To demonstrate that there was indeed energy transfer between the CFP and YFP in our FRET sensor, we measured the emission spectra of the purified sensor protein using a spectrofluorometer. When we excited the sensor C3 protein with light at the excitation wavelength for CFP (433 nm), the majority of the fluorescence light was detected at the emission wavelength of YFP (526 nm), indicating that there was an efficient energy transfer between the donor molecule (CFP) and the acceptor molecule (YFP) in sensor C3 (Figure 5A). We then examined whether the FRET effect was sensitive to caspase-3 cleavage. We mixed 1.28 µg of purified sensor C3 protein with different amounts of caspase-3 for 1 hr and then measured its emission spectra using a spectrofluorometer. The emission of YFP decreased steadily with increased amount of caspase-3. When the caspase concentration reached 20 nM, the emitted light was mainly from CFP (480 nm) (Figure 5A).

To determine the sensitivity of sensor C3 on the caspase-3 activity, we have performed a series of dosage-dependent experiments similar to the one shown in Figure 5A. The analyzed results are shown in Figure 5B, in which the relative emission ratio of YFP/CFP (measured at 526 nm / 480 nm) was plotted against the concentration of caspase-3 in a log scale. Here, the emission ratio (YFP/CFP) was normalized against that of the un-cleaved sensor C3 protein. Based on these data, it is clear that the FRET effect of the sensor C3 protein is highly sensitive to the activity of caspase-3.

But, we still need to test that, under a physiological condition (i.e. during the process of apoptosis in a living cell), whether the cleavage of this probe is fast enough so that one can measure the dynamics of caspase-3 activation based on the changes of FRET. In order to achieve this goal, we must first determine what is the concentration of caspase-3 in a cell undergoing apoptotic cell death and what is the concentration of sensor C3 when it is expressed in a cultured mammalian cell.

To measure the concentration of active caspase-3 in an apoptotic cell, we first generated a standard curve by performing a series of caspase-3 assays using known amounts of caspase-3 protein and excessive amount of fluorescent substrate, Ac-DEVD-AMC. Then, HeLa cells were induced into apoptosis by UV-irradiation. Six hours after UV treatment, cell lysates were collected and their caspase-3 activities were determined. Based on a comparison with the standard curve, we estimated that the average concentration of active caspase-3 within an apoptotic HeLa cell was about 0.35 µM. To estimate the amount of sensor protein presented in a HeLa cell, we generated another calibration curve by using a confocal microscope to measure the fluorescent signal of YFP in small protein droplets (embedded in oil) containing known concentrations of the purified sensor C3 protein. Then, YFP images of HeLa cells (n=205) expressing sensor C3 were recorded under identical imaging condition. By comparing the fluorescent intensities of YFP measured from the sensor C3-expressing cells with the calibration curve, we estimated that the average concentration of sensor C3 in an

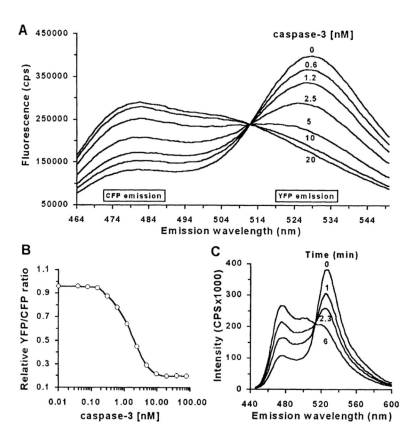

Figure 5. (A) Results of a spectrofluorometer analysis of the purified sensor C3 protein. The excitation wavelength was 433 nm. Samples were pre-incubated with various concentrations of caspase-3 for one hour. (B) Based on spectrofluorometer analysis similar to those shown in panel A, the emission ratio between YFP (526 nm) and CFP (480 nm) was determined as a function of the caspase-3 concentration. (C) The purified sensor C3 protein (0.14 µM) was incubated with caspase-3 (0.01 µM) at 37°C for different period of time. At each indicated time point, the emission spectra of sensor C3 was measured using a spectro-fluorometer. The FRET changes indicated cleavage of sensor C3 by caspase-3.

individual cell was about 5.0 μM. Thus, when a HeLa cell is undergoing apoptosis, the molar ratio between the expressed sensor C3 and the active caspase-3 is about 14 to 1.

We then performed a caspase-3 cleavage assay by adding 0.01 μM of caspase-3 into 0.14 μM of sensor C3 (which has the same *in vivo* ratio of caspase-3 to sensor) and conducted a time-dependent emission scanning analysis on the reaction mixture using a spectrofluorometer. The results are shown in Figure 5C. At 0 min, which is prior to the addition of caspase-3 to the sensor protein, a large emission peak was detected from YFP (526 nm) when sensor C3 was excited at the excitation wavelength of CFP (433 nm), while a much smaller peak was detected from CFP (476 nm). The intensity of YFP emission was quickly reduced 1 min after the caspase-3 addition and reached to the lowest level in 6 min. Meanwhile, the intensity of CFP emission increased to the maximum level. This result shows that under the *in vivo* ratio of caspase-3 and sensor C3, caspase-3 can cleave most of the sensor C3 molecules within several minutes. Thus, one can measure the dynamics of caspase-3 activation during apoptosis in a living cell based on the changes of FRET.

3.4. Application of Sensor C3 to Measure the Dynamic Activation of Caspase-3 in a Single Living Cell During Apoptosis

To demonstrate the usefulness of this FRET biosensor for studying caspase-3 activity in living cells, we have introduced the gene encoding sensor C3 into HeLa cells. After the gene was expressed, the cells were induced to enter apoptosis by UV-irradiation. To measure the FRET effect under the *in vivo* condition, we recorded the fluorescence images of the cells using a digital imaging system. We excited the sensor C3 probes in the cells using excitation light of CFP (440±10 nm), and recorded two separated fluorescent images of the cells using the emission filters of CFP (480±15 nm) and YFP (535±13 nm). By observing the intensity shift between the YFP and CFP images, one can detect the change of the FRET effect in response to the cleavage of the probe. Sample results of such a measurement are shown in Figure 6A. Here, two neighboring cells expressing the sensor C3 gene were undergoing apoptosis. (Their phase images are displayed in panel a).

We merged the CFP image (colored blue) and the YFP image (colored green) of the same cells together and displayed it in panels of b-k. It is evident that, before caspase-3 was activated, the cells were in the normal cell shape and the fluorescence light was emitted primarily from the YFP (panels b – f). (Hence, the images were predominantly green). As caspase-3 was activated during apoptosis, sensor C3 was cleaved and the fluorescence light was emitted mainly from the CFP due to the abolishing of FRET. As a result, the merged image became blue (panels g – k). Using this method, one can easily detect the activation of caspase-3 inside a living cell by observing the shift of color in the merged fluorescence image. From Figure 6A, it can be seen that the two neighboring cells initiated their caspase-3 activation at slightly different times.

One can also see that, in both cells, the activation of caspase-3 took place far before any cell morphological change was observed. In addition, our FRET probe can be used to provide a quantitative measurement of the activity of caspase-3 in a living cell. As shown in Figure 5B, the intensity ratio between YFP and CFP emissions was related to the caspase-3 activity. Such a ratio can be determined in the living cell by dividing the pixel value of the YFP image (in a chosen area of the cell) with that of the CFP image, under

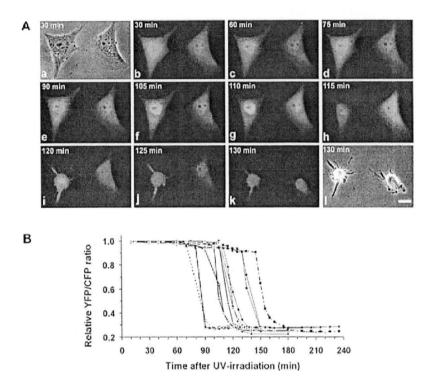

Figure 6. (A) Dynamic FRET changes of sensor C3 in response to caspase-3 activation during UV-induced apoptosis in living HeLa cells. The phase images (panels a and l) and the fluorescence images of YFP and CFP (panels b-k) of two neighboring cells were recorded by a cooled CCD camera. (The time after UV treatment was indicated in each panel). The CFP image (colored blue) was merged with the YFP image (colored green) (panel b-k). The cells were excited at the excitation wavelength of CFP. A reduction of the FRET effect was represented by a shift of color from green (i.e., YFP emission) to blue (i.e., CFP emission). Bar, 10 μm. (B) Dynamic changes of the FRET effect of sensor C3 during UV-induced apoptosis measured from 11 different HeLa cells. Results were obtained from living cell measurements similar to those shown in Figure 6A. The relative emission ratio of YFP/CFP (which was normalized to 1.0 at t = 0) was plotted as a function of time. There was a dramatic decrease of this ratio when the probe was cleaved by caspase-3.

the condition that only the CFP portion of the probe was excited. Using this method, we have determined the dynamics of caspase-3 activation in a number of apoptotic cells. The results from 11 different cells are summarized in Figure 6B. Here, the relative emission ratio of YFP/CFP was plotted as a function of time. (The YFP/CFP ratio of the intact probe was normalized as 1.0). There were more than 4-fold decreases when the probe was cleaved by the caspase-3. It is evident that each individual cell initiated its caspase-3 activation at very different times. The average time of the onset of caspase-3 activation was 107 ± 29 min (n = 20) after the UV-treatment. Yet, the time required to cleave the sensor C3 probe within the cell was relatively uniform among different cells (18 ± 4 min, n = 20). These results demonstrated that this FRET probe is highly useful for assaying the dynamics of caspase-3 activation in living cells during apoptosis.

3.5. Advantages of the FRET-based Bio-sensor

In summary, based on results of the above studies, our FRET-based biosensor C3 seems to have a number of advantages, including:

(1) This sensor C3 not only gives a strong fluorescence signal, it also has a large FRET effect. Our *in vivo* study indicated that the change of emission ratio (YFP/CFP) of our probe in response to caspase-3 activation in a living cell was 4.08 fold (n = 13), which was significantly larger than the value observed in most other earlier FRET studies (which was typically 2 fold or less) (Tyas et al., 2000). Up to this point, sensor C3 is the most sensitive FRET probe developed for assaying caspase activation in living cells.
(2) This probe is a very efficient substrate of caspase-3. As demonstrated in Figures 5 and 6, our probe can be easily cleaved by caspase-3 activated within a single cell. This is mainly due to the special design of the linkers flanking the DEVD substrate site, which allows an easy access of caspase to the cleavage site.
(3) Because of the rapid cleavage of this probe, sensor C3 can be used effectively to measure the dynamics of activation of the endogenous caspase-3 within a cell. Such measurement would be very difficult to do using other methods.

We believe that this FRET probe is a very useful technology for both basic research and drug discovery. First, since this bio-sensor is a gene product, it can be automatically produced by the transfected cells. Stable cell lines containing the molecular bio-sensor can also be generated. In fact, we have generated a stable cell line of sensor C3 and used these cells to detect caspase-3 activation during UV and TNF-α induced apoptosis. Second, this FRET-based sensor can be used for a variety of apoptotic induction treatments. Third, it can be an important tool for studying the signal transduction mechanisms of cell death. It is well known that the activation of caspase-3 is the most critical event in apoptosis since it marks the beginning of the execution phase for programmed cell death. Using this FRET method, one can easily correlate the various apoptotic events with the event of caspase-3 activation within a single living cell. Forth, using this FRET probe, apoptosis can be assayed by optical detection; it is fast and simple. Thus, this method is highly useful for large-scale drug screening or toxicity testing. Many important diseases, including cancer, stroke, neurodegenerative disease and auto-immune diseases, are related to defective or excessive programmed cell death. Thus, drugs that can either facilitate or block programmed cell death can both be potentially useful in treating diseases. This FRET based bio-sensor can be used as a platform technology for drug discovery by screening active chemical components in traditional Chinese medicine (TCM) that are found to interfere with the apoptotic process.

4. ACKNOWLEDGEMENT

This work was supported by the Research Grants Council of Hong Kong SAR (HKUST6109/01M and HKUST6104/02M) and the HIA project of HKUST.

5. RERERENCE

Chalfie, M., Tu, Y., Euskirchen, G., Ward, W.W. and Prasher, D.C. 1994, Green fluorescent protein as a marker for gene expression. *Science*, **263**: 802-805.

Cohen, G.M. 1997, Caspases: the executioners of apoptosis. *Biochem J*, **326**: 1-16.

Desagher, S. and Martinou, J.C. 2000, Mitochondria as the central control point of apoptosis. *Trends Cell Biol*, **10**: 369-377.

Du, C., Fang, M., Li, Y., Li, L. and Wang, X. 2000, Smac, a mitochondrial protein that promotes cytochrome c-dependent caspase activation by eliminating IAP inhibition. *Cell*, **102**: 33-42.

Forster, T. 1959, Transfer mechanisms of electronic excitation. *Discuss. Farraday Soc.*, **264**: 7-17.

Gao, G. and Dou, Q.P. 2000, N-terminal cleavage of bax by calpain generates a potent proapoptotic 18-kDa fragment that promotes bcl-2-independent cytochrome C release and apoptotic cell death. *J Cell Biochem*, **80**: 53-72.

Gao, W., Pu, Y., Luo, K.Q. and Chang, D.C. 2001, Temporal relationship between cytochrome c release and mitochondrial swelling during UV-induced apoptosis in living HeLa cells. *J Cell Sci*, **114**: 2855-2862.

Heim, R. and Tsien, R.Y. 1996, Engineering green fluorescent protein for improved brightness, longer wavelengths and fluorescence resonance energy transfer. *Curr Biol*, **6**: 178-182.

Jacobson, M.D., Weil, M. and Raff, M.C. 1997, Programmed cell death in animal development. *Cell*, **88**: 347-354.

Luo, K.Q., Yu, V.C., Pu, Y. and Chang, D.C. 2001, Application of the fluorescence resonance energy transfer method for studying the dynamics of caspase-3 activation during UV-induced apoptosis in living HeLa cells. *Biochem Biophys Res Commun*, **283**: 1054-1060.

Martin, S.J. and Cotter, T.G. 1991, Ultraviolet B irradiation of human leukaemia HL-60 cells in vitro induces apoptosis. *Int J Radiat Biol*, **59**: 1001-1016.

Nagata, S. 1997, Apoptosis by death factor. *Cell*, **88**: 355-365.

Oren, M. and Rotter, V. 1999, Introduction: p53--the first twenty years. *Cell Mol Life Sci*, **55**: 9-11.

Reed, J.C., Jurgensmeier, J.M. and Matsuyama, S. 1998, Bcl-2 family proteins and mitochondria. *Biochim Biophys Acta*, **1366**: 127-137.

Shimomura, O., Johnson, F.H. and Saiga, Y. 1962, Extraction, purification and properties of aequorin, a bioluminescent protein from the luminous hydromedusan, Aequorea. *J Cell Comp Physiol*, **59**: 223-239.

Thompson, C.B. 1995, Apoptosis in the pathogenesis and treatment of disease. *Science*, **267**: 1456-1462.

Tyas, L., Brophy, V.A., Pope, A., Rivett, A.J. and Tavare, J.M. 2000, Rapid caspase-3 activation during apoptosis revealed using fluorescence-resonance energy transfer. *EMBO rep*, **1**: 266-270.

Wang, X. 2001, The expanding role of mitochondria in apoptosis. *Genes & Development*, **15**: 2922-2933.

Yang, F., Moss, L.G. and Phillips, G.N., Jr. 1996, The molecular structure of green fluorescent protein. *Nat Biotechnol*, **14**: 1246-1251.

Zhang, J., Campbell, R.E., Ting, A.Y. and Tsien, R.Y. 2002, Creating new fluorescent probes for cell biology. *Nat Rev Mol Cell Biol*, **3**: 906-918.

Chapter 4

FLUORESCENCE RESONANCE ENERGY TRANSFER (FRET) STUDY ON PROTEIN-PROTEIN INTERACTION IN SINGLE LIVING CELLS

Xun Shen[1], Chunlei Zheng[1], Ziyang Lin[2], Yajun Yang[1] and Hanben Niu[2]

1. INTRODUCTION

The cell can be conceived as a biochemical information-processing device whose response to the environment depends on the state of spatially organized networks of protein activities. The interconnectivity and spatial organization of protein systems that sustain basic cellular functions is only maintained in the context of the whole intact molecular architecture of the cell. It is clear that understanding cell function by integrating molecular activities within the living cell is a big challenge for modern biology. Numerous biochemical assays revealing protein modifications, interactions or transport have already helped to bring us closer to this goal. However, none of the methods that are based on reconstituted systems in the test-tube can fully take into account the compartmentalized and interconnected nature of these reactions in cells. It is therefore necessary to develop methods that can measure the dynamics of these biochemical reactions in the intact cell and thereby extract the spatial organization *in vivo*. Only recently, and most probably because of the availability of genetically encoded fluorescent proteins, it has become possible to image not only cellular processes such as protein translocation or transport but also basic reactions such as protein interactions, proteolysis and phosphorylation in intact cells. In this chapter, we will introduce a newly developed technology, which is called fluorescence resonance energy transfer, for studying protein-protein interaction in single living cell through an example in which the interaction of the small heat shock protein 27 (hsp27) with p38, the member of the

---

[1] Institute of Biophysics, Chinese Academy of Sciences, Beijing 100101, China. shenxun@sun5.ibp.ac.cn

[2] Institute of Optoelectronics, Key Laboratory of Opto-electronics Devices and Systems, Shenzhen University, Ministry of Education, Shenzhen 518060, China.

mitogen-activated protein kinases (MAPK) family is concerned.

The current methods to investigate protein interaction mainly include yeast two-hybrid system (Luban and Goff, 1995) and co-immunoprecipitation (Anderson, 1998). The former employs a transcriptional 'read-out' in yeast and relies on the discovery that virtually any pair of proteins that interact with each other may be used to bring separate activation and DNA-binding domains together to reconstitute a transcriptional activator if one could be fused to a specific DNA-binding domain and the other to a transcriptional activation domain (Fields and Song, 1989). However, the requirement for stable expression of the fusion proteins and that the fusion proteins must be capable of transport to the nucleus seriously limits the application of this method. In co-immunoprecipitation study, the agarose beads coated with a selective antibody to precipitate a known protein from crude cell extract, then other specific antibodies are used to identify if any other unknown protein is co-precipitated with the known protein. Thus, if protein X forms a stable complex with protein Y in cell, immunoprecipitation of X may result in co-precipitation. However, the study of protein-protein interaction with this method can only be performed in vitro. Non-specific binding of two proteins cannot be excluded, and the low-affinity or transient binding of proteins is hardly to be detected. Besides those, co-precipitation can not applied to study the interaction involving insoluble protein.

The 'green revolution' initiated by the introduction of the green fluorescent protein GFP from Aequorea victoria (Chalfie et al., 1994) and the later developments of GFP-mutants possessing different spectral properties (Pollok and Heim, 1999) offered the possibility of simultaneous expression of donor and acceptor fusion proteins in the same cell and allowed measurement of their interactions by fluorescence resonance energy transfer (Miyawaki, 1997; Mitseli and Spector, 1997). At present, the combination of CFP (cyan fluorescence protein) donor and YFP (yellow fluorescence protein) acceptor fusion proteins are particularly useful. This effective FRET pair can be used to monitor the proximity of the two attached fluorescent proteins in 3-6 nm (Tsien, 1998). Co-expression of CFP- and YFP-fusion proteins has been successfully used to analyze short-time changes in protein-protein interactions, e.g. oligomerization, colocalization, complex formation (Ellenberg et al, 1999), activation of protein kinases (Ng et al. 1999) and mapping of enzyme activities in living cells (Bastiaens and Pepperkok, 2000).

2. PRINCIPLE OF FLUORESCENCE RESONANCE ENERGY TRANSFER (FRET)

FRET was first described by Förster (1948). It has become extremely important for modern cell biology, because FRET allows measuring distances between molecules on a scale of a few nanometers. This is far below the resolution limit of modern optical far field microscopy, which currently is at approximately 100 nm. Forster's theory explains FRET as a dipole-dipole interaction between neighboring molecules and derives the dependence of the energy transfer efficiency E on their actual proximity R. For efficient FRET the distance R between these two molecules, the excited donor D and the fluorescent acceptor A, is typically 2-7 nm. The direct non linear dependence between E and R can be described as

$$E = 1 / [1+(R/R_0)^6] \qquad (1)$$

where R_0 is the distance between the D and A for $E=0.5$. R_0 reflects the properties of a particular D-A pair including acceptor quantum yield, spectral overlap between D-emission and A-excitation and the relative spatial D-A orientation.

The extreme sensitivity of the FRET process on the distance between molecules 6th power dependence on R renders it a very useful tool for the resolution of intracellular arrangement and dynamics of biological molecules. However, fluorophore bleaching and induction of cell damage limit the application of the FRET technique for the study of long-term (minutes to hours) processes of transport or signal transduction in living cells. The problem can be ameliorated by applying two-photon excitation instead of one-photon excitation. Since FRET occurs over distances similar to the size of proteins, it can be used to extend the resolution of the fluorescence microscope (typically 250 nm) to detect protein-protein interactions. FRET microscopy is thus an ideal technique to determine whether proteins that are co-localized at the level of light microscopy interact with one another in cells.

The CFP and YFP are most widely used donor and acceptor chromophores for construction of the fused protein(s) containing the investigated protein(s) in FRET study. Currently, there are two types FRET: intramolecular and intermolecular FRET. In intramolecular FRET, both the donor and acceptor chromophores are on the same host molecule, which undergoes a conformation transition or as substrate for an investigated enzyme. Any conformational change or cleavage of the molecule may yield a large change or even lose of the FRET. In intermolecular FRET, one investigated molecule (protein A) fused to the donor (CFP) and another molecule (protein B) fused to the acceptor (YFP). When the two proteins bind to each other, FRET occurs. When they dissociate, FRET diminishes. In FRET experiments, a single transfection (intramolecular FRET) or co-transfection (intermolecular FRET) of the constructs must first be performed. The occurrence of FRET can be observed by exciting the sample at the donor excitation wavelengths while measuring the fluorescence intensities emitted at wavelengths corresponding to the emission peaks of the donor versus those of the acceptor. If the acceptor and donor are at a favorable distance and orientation, donor emission intensity decreases (CFP, cyan) while the acceptor emission (YFP, yellow)

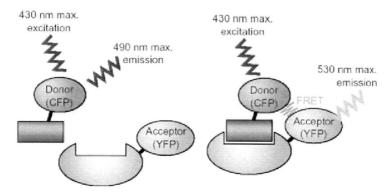

Figure 1. The principle of intermolecular FRET for studying protein-protein interaction. (cited from van Roessel, P and Brand)

intensity increases. Figure 1 depicts the fluorescence intensity changes of the donor and acceptor before and after FRET takes place.

3. FRET MICROSCOPY: METHODS FOR FRET MEASUREMENT

FRET results in several measurable phenomena, including sensitized acceptor fluorescence, quenching of donor fluorescence, and decrease in donor fluorescence lifetime. There are many ways to measure FRET in a microscope. These methods can be divided into intensity-based methods and lifetime-based methods.

3.1. Sensitized Acceptor Fluorescence

The fluorescence from the acceptor is measured while exciting the donor. This is typically done using three sets of filters which consist of CFP channel (excited at the excitation wavelength of CFP and measuring fluorescence at the emission wavelength of CFP), YFP channel (excited at the excitation wavelength of YFP and measuring fluorescence at the emission wavelength of YFP) and FRET channel (excited at the excitation wavelength of CFP and measuring fluorescence at the emission wavelength of YFP). However, the net FRET signal must be corrected against bleed-through of the non-FRET CFP fluorescence and the YFP fluorescence due to excitation at the excitation wavelength of CFP but not to the energy transfer from excited CFP according to the formula:

$$\text{Net FRET signal} = \text{FRET signal} - a \cdot \text{YFP signal} - b \cdot \text{CFP signal} \quad (2)$$

where a and b are the ratio of the signal in FRET channel to the signal in YFP channel in the absence of donor and to the signal in CFP channel in the absence of acceptor respectively. The coefficient a and b can be obtained from the measurement in the cells transfected with CFP fusion protein or YFP fusion protein alone. Obviously, a positive net FRET signal detected in the cells transffected with both fusion proteins indicates a fluorescence resonance energy transfer from donor to acceptor.

3.2. Acceptor Photobleaching Approach

FRET can be also accomplished by comparing the donor fluorescence intensity in the cells transfected with both CFP fusion protein (donor) and YFP fusion protein (acceptor) before and after destroying the acceptor by photobleaching. If FRET is initially present, fluorescent intensity of donor will increase after acceptor is photobleached. Energy transfer efficiency, E, can be calculated according to following formula:

$$E = 1 - F_{DA} / F_D \quad (3)$$

with F_{DA} the fluorescence intensity of the donor in the presence of the acceptor and F_D the fluorescence intensity of the donor after acceptor is photobleached. An advantage of this method is that it requires only a single sample and that the energy transfer efficiency can thus be directly correlated with donor fluorescence before and after photobleaching of the acceptor.

3.3. Fluorescence Lifetime Imaging Microscopy (FLIM)

An alternative method is fluorescence lifetime imaging. Fluorescence lifetime of the donor is reduced by FRET and this effect can be directly measured by fluorescence lifetime imaging microscopy (FLIM) (Bastiaens and Squire, 1999). Fluorescence lifetime imaging microscopy is a technique in which the mean fluorescence lifetime of a chromophore is measured at each spatially resolvable element of a microscope image. The nanoseconds excited-state lifetime is independent of the chromophore concentration or light path length but dependent upon the excited-state reaction such as fluorescence resonance energy transfer. Imaging using fluorescence lifetimes may also provide functional data about the protein being probed since the lifetime of a fluorophore can be a function of its microenvironment within cell. Fluorescence lifetime imaging can be achieved by frequency domain techniques or by time-domain techniques. Frequency domain techniques use the phase shift between the modulated or pulsed excitation and the emission of the sample at the fundamental modulation frequency or its harmonics. Lifetime imaging can be achieved by modulated image intensifiers and wide-field microscopes (Squire et al., 2000) or by modulating single channel detectors used in laser scanning microscopes (Carlsson and Liljeborg, 1998). Time-domain techniques use pulsed excitation and record the fluorescence decay function directly. Lifetime imaging is achieved by gated image intensifiers (Cole et al., 2001), by directly gated CCDs (Mitchell et al., 2002), by counting the photons in several parallel time gates (Sytsma et al., 1998) or by time-correlated photon counting (Becker et al., 2004).

Instead of measuring emission intensity, the fluorescence lifetime of the donor alone (τ_D) and also in the presence of the acceptor (τ_{DA}) are measured. If FRET occurs, τ_{DA} will be different from τ_D and this difference can be used to calculate FRET efficiency E as

$$E = 1 - \tau_{DA}/\tau_D \qquad (4)$$

The major advantage of FLIM is that it permits an internally calibrated measurement of FRET. Also, as only donor emission is monitored, factors that affect the quantum yield of the acceptor can be disregarded.

4. STUDIES ON THE INTERACTION BETWEEN HEAT SHOCK PROTEIN 27 AND P38 MAP KINASE

4.1. Small Heat Shock Protein 27 and p38 MAP Kinase

The 27 kDa stress response small heat shock protein 27, a marker of differentiation and proliferation, helps the cell in repair processes after environmental stress such as heat, UV-irradiation and oxidative stress. It has shown activation-dependent translocation from the cytosolic to the nuclear region and has been linked to the cellular stress response. It plays many other roles in the regulation of cell function, such as inhibiting death receptor-mediated apoptosis (Ricci et al., 2001), promoting growth of human astrocytomas (Khalid et al., 1995), characterizing the tumor as relatively benign and slow progressing (Bayerl et al., 1999). p38 (the α-isoform) is a member of mitogen-activated protein kinase (MAPK) family and often activated by stress and various cytokines. It contains the phosphoacceptor sequence Thr-gly-tyr. In the signaling cascade, it is activated via phosphorylation of Thr and Tyr residues by a dual-specificity serine-threonine MAPK-kinase (MKK), and then phosphorylates and activates MAPK-activated protein kinase-2 (MK2). The later then phosphorylates hsp27 (Freshney et al., 1994; Rouse et al., 1994). It was reported that when shock protein hsp27 became phosphorylated, the intracellular distribution of hsp27 was changed from the cytoplasm to the peri-nuclear region (Nakatsue et al., 1998). However, it has not been known if there is any direct interaction between hsp27 and p38. In this study, by using FRET technology we demonstrated that p38 could directly interact with hsp27 and the interaction depended on the activation of p38 and phosphatidylinositol 3-kinase (PI3K).

4.2. Experimental Procedures

4.2.1. Expression Vectors and Cell Transfection

Hsp27 cDNA and p38 cDNA were obtained by PCR amplification from vector, pBluescript-hsp27, and pCDNA3-Flag-p38 respectively. The CFP-hsp27 fusion construct was made by subcloning Hsp27 cDNA into cyan fluorescence protein (CFP) expressing vector, pECFP-C1, in EcoRI-BamHI sites. The p38 cDNA was subcloned into yellow fluorescence protein (YFP) expressing vector, pEYFP-C1, in HindIII-XhoI sites to construct the YFP-p38 fusion protein-expression vector. A glycine was inserted between YFP and p38 as linker in the YFP-p38 fluorescence chimera.

The mouse fibroblast cells (L929 cell line) were planted on glass–bottomed dishes (MatTek Corp.) and transfected with either CFP-hsp27 plasmid or YFP-p38 plasmid DNA or both using Lipofectamine 2000. 12 h after transfection, the full-length proteins (CFP-hsp27 or/and YFP-p38) expressed from each chimera can be observed in some cells.

4.2.2. Microscopy

As shown in Figure2, all FRET microscopic observations were performed on a Leica DM IRE2 confocal laser scanning microscope system at 37 °C 12 h after transfection. Excitation was provided by multimode argon ion laser beam using a double 458 nm/514 nm diachronic splitter. Donor (CFP) was excited at 458 nm and its fluorescence was detected in a bandwidth of 478-498 nm (CFP channel), whereas the excitation at 514 nm and emission at 545 ± 15 nm were used for detecting acceptor (YFP) (YFP channel). For FRET, the excitation was at 458 nm and detection at 545 ± 15 nm (FRET channel). The fluorescence images of the transfected cells were taken up at CFP-, YFP- and FRET-channel respectively. The FRET signal was corrected against bleed-through of the non-FRET CFP and YFP fluorescence.

Time-correlated single photon counting (TCSPC) fluorescence lifetime imaging using a multi-photon confocal laser scanning microscope system (Leica DM IRE2 and Becker & Hickl SPC730) was performed to obtain the donor CFP fluorescence lifetime images in the cells transfected with CFP-hsp27 chimera alone and co-transfected with both CFP-hsp27 and YFP-p38. The femto-seconds pulsed laser (Coherent Mire 900) beam was used to excite donor CFP in the transfected cells. The laser power was adjusted to give an average photon-counting rate of 10^4-10^5 photons·s^{-1}. Cells were imaged for 50s to achieve appropriate photon statistics for determination of the fluorescence dynamics.

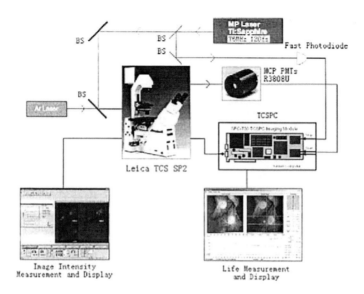

Figure 2. The fluorescence microscopy imaging system for FRET studies. It consists of a Leica TCS SP2 confocal laser scanning microscope, a time-correlated single photon counting unit (TCSPC), an argon ion laser, a femto-seconds pulsed laser (76 MHz, 120 fs) and two displayers for image intensity and lifetime measurement respectively.

4.3 Results

4.3.1. Interaction of p38 and Hsp27 in Quiescent Cell

As shown in Figure3, the fluorescence from CFP and YFP can be observed simultaneously in the cell transfected with both CFP-hsp27 and YFP-p38 fusion chimeras. It can be seen that the heat shock proteins (hsp27) are distributed in cytoplasm, while the p38 MAP kinase in both cytoplasm and nuclei. The net fluorescence signal observed in FRET channel indicates the existence of an energy transfer from CFP to YFP.

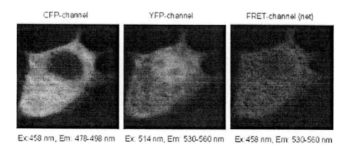

Figure 3. Fluorescence images of the cell transfected with both CFP-hsp27 and YFP-p38 detected in CFP-, YFP- and FRET-channels respectively.

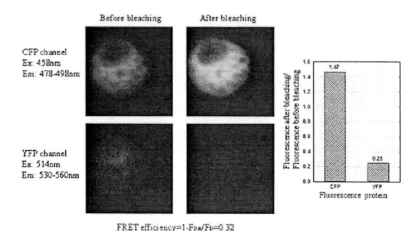

Figure 4. The fluorescence intensity images of CFP-hsp27 and YFP-p38 in the L929 cell expressing the two fusion proteins before and after photobleaching of YFP.

In order to determine the FRET efficiency, dequenching of the donor CFP by selective photobleaching of the acceptor YFP was performed. In the experiments, the transfected cells were illuminated at the YFP excitation wavelength (514 nm) for 2 min at full laser power, and then CFP-hsp27 images were taken up at the same focal plane. To quantify changes in the CFP and YFP fluorescence intensity before and after bleaching, selected regions of the images were quantified using Leica confocal software. As shown in Figure4, the fluorescence intensity of CFP significantly increased after photobleaching of YFP. Based on about 47% increase in the fluorescence intensity of CFP after photobleaching the acceptor YFP, the FRET efficiency of 32% was found between these two fusion proteins.

The FRET between these two fusion proteins was also studied by fluorescence lifetime imaging microscopy. The fluorescence intensity images and lifetime images of CFP in the cells transfected with CFP-hsp27 alone or with CFP-hsp27 and YFP-p38 together is shown in Figure5. It is difficult to judge whether the fluorescence intensity of CFP in the
cell expressing both the CFP-hsp27 andYFP-p38 fusion protein is less than that in the cell expressing the CFP-hsp27 fusion protein alone only by intensity measurements. However, the lifetime imaging clearly shows the shortening of the fluorescence lifetime of CFP when the CFP-hsp27 fusion protein is co-expressed with YFP-p38 fusion protein in cell. The FLIM measurement seems much better than fluorescence intensity measurement for showing the interaction between hsp27 and p38.

4.3.2. Interaction of p38 with Hsp27 in H_2O_2-stimulated Cell

The results in previous paragraph only show existence of the interaction of p38 with hsp27 when p38 MAP kinase is not activated. In order to know if the interaction

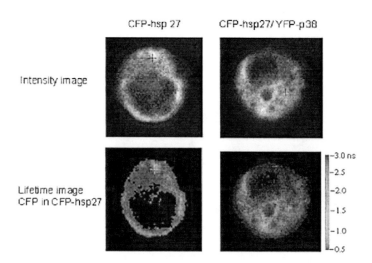

Figure 5. Fluorescence intensity and lifetime images of CFP in the L929 cell expressing CFP-hsp27 alone (left panels) or expressing both CFP-hsp27 and YFP-p38 (right panels). In fluorescence lifetime images (lower panels) different color represent different lifetime (ns).

depends on the activity or phosphorylation of the MAP kinase, hydrogen peroxide was used to stimulate the cell co-transfected with the CFP-hsp27 fusion protein and YFP-p38 fusion protein. The H_2O_2-stimulation will result in an activation or phosphorylation of p38. Thus, the FRET study on the stimulated transfected cells may tell us more information on the regulation of the interaction. At first, the fluorescence intensity imaging approach was performed with the cell expressed both fusion proteins. 1 mM H_2O_2 was used to stimulate the cells, intensity images before stimulation and at various time after the stimulation were taken in CFP-channel and FRET-channel simultaneously. The results are shown in Figure6. It can be seen that after stimulation the fluorescence intensity in CFP-channel became increasing while the intensity in FRET-channel was decreasing. The results suggested that the interaction between p38 and hsp27 becomes weakening upon activation of p38 MAP kinase.

Same as seen in the FLIM results in study of FRET between the two fusion proteins in quiescent cells, the FLIM approach is proved to be much better for studying the dynamic process of the interaction between the CFP-hsp27 and the YFP-p38 fusion proteins in the stimulated cells. Figure7 shows the fluorescence intensity images and lifetime images of CFP in the cell expressing both fusion proteins at various moments after the stimulation by H_2O_2. It is very straightforward to see the dynamic change of the interaction as a gradual increase of the fluorescence lifetime of CFP after stimulation. It was observed that the average fluorescence lifetime of CFP increased from 1.17 ns before H_2O_2-stimulation to 1.34, 1.52 and 1.74 ns at 6, 12 and 18 min after the stimulation. The results suggest that the interaction of p38 with hsp27 is getting weak and weak as p38 MAP kinase being activated by H_2O_2. The results clearly demonstrate that activation or phosphorulation of p38 will lead to diminish of the interaction between p38 and hsp27.

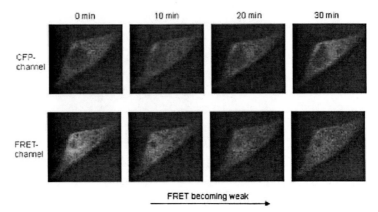

Figure 6. Increasing of the fluorescence intensity in CFP-channel and decreasing of the fluorescence intensity in FRET-channel in the L929 cells expressing both CFP-hsp27 and YFP-p38 fusion proteins after 1 mM H_2O_2-stimulation.

To further testify the p38 activity-dependent interaction between p38 and hsp27, a selective inhibitor of p38, SB203580, was used to treat the cells before stimulation by H_2O_2. Figure8 shows the observed fluorescence intensity images and lifetime images of CFP in the cell, which has two fusion proteins expressed and has been pre-incubated with 2 μM SB203580 for 20 min before stimulation, at various moments after the stimulation by H_2O_2. It was found that p38 inhibitor prevent the lifetime of CFP from increase when cell was stimulated by H_2O_2. This indicates that if p38 is unable to be activated, H_2O_2-stimulation would not affect the interaction between hsp27 and p38.

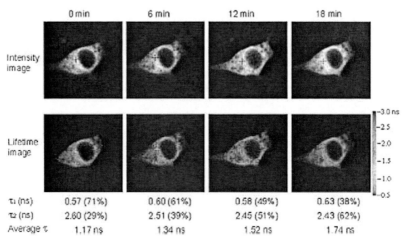

Figure 7 The fluorescence intensity images (up panels) and the fluorescent lifetime images (low panels) of CFP in the L929 cell simultaneously expressing the CFP-hsp27 and YFP-p38 fusion proteins at various time after stimulation by 1 mM H_2O_2.

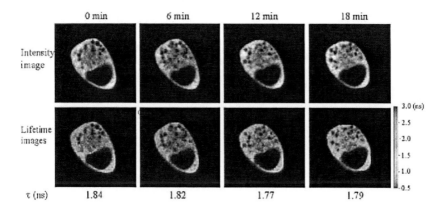

Figure 8. The fluorescence intensity images (up panels) and the fluorescent lifetime images (lower panels) of CFP in the L929 cell, which has both the CFP-hsp27 and YFP-p38 fusion proteins expressed and has been pre-incubated with 2μM SB203580 for 20 min before stimulation, after stimulation by 1 mM H_2O_2.

4.3.3. Interaction of p38 with hsp27 in the Cell Stimulated by Arachidonic Acid

Since the serine/threonine kinase protein kinase B, also called Akt, is found to exist in a signaling complex containing p38 kinase, MK2 and hsp27 (Rane et al., 2001), Akt may participate in the regulation of the interaction of p38 with hsp27. It has also been reported that Akt activation is dependent on phosphatidylinositol 3-kinase (PI3K) (Burgering and Coffer, 1995) and PI3K can be activated by arachidonic acid in a variety of cell type (Hii et al., 2001). Thus, arachidonic acid was use to stimulate the L929 cells transfected with both CFP-hsp27 and YFP-p38 fusion protein expressing vectors, and to see if the FRET between these two fusion proteins was affected by activation of Akt. The fluorescence intensity images and lifetime images of CFP in the cell expressing both fusion proteins were taken at various moments after the stimulation by 10 μM arachidonic acid. As sown in Figure9, the fluorescence lifetime of CFP in the cell increased from 1.69 ns before arachidonic acid-stimulation to 1.9, 2.05 and 2.11 ns at 6, 12 and 18 min respectively after the stimulation, which means that the FRET becomes weaker and weaker after the stimulation. The results may suggest that the activation of Akt leads to loss of the interaction of p38 with hsp27.

In order to further testify the role of Akt in regulating the interaction of p38 with hsp27, wortmannin, an inhibitor of phosphatidylinositol 3-kinase (PI3K) upstream of Akt, was used to inactivate the activation of Akt in arachidonic acid-stimulated cells. The fluorescence intensity images and lifetime images of CFP in the cell expressing the two fusion proteins were taken at various moments after stimulation by arachidonic acid in the presence of 100 nM wortmannin. The results are shown in Figure10. It is evident that the presence of wortmannin prevents the lifetime of the CFP-hsp27 from being lengthening in the cells expressing both CFP-hsp27 and YFP-p38 after the stimulation. The results suggests that inactivation of Akt by the inhibitor of PI3K also prevent the interaction between p38 and hsp27 from being lost.

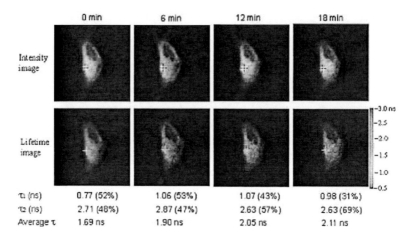

Figure 9. The fluorescence intensity images (up panels) and the fluorescent lifetime images (lower panels) of CFP in the L929 cell expressing both the CFP-hsp27 and YFP-p38 fusion proteins before and after stimulation by 10 μM arachidonic acid.

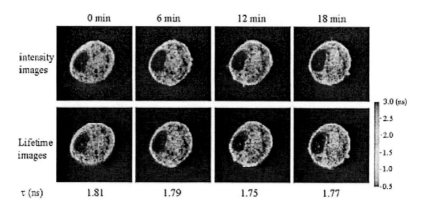

Figure 10. The fluorescence intensity images (up panels) and the fluorescent lifetime images (lower panels) of CFP in the L929 cell expressing both the CFP-hsp27 and YFP-p38 fusion proteins and in the presence of 100 nM wortmannin, the PI3K inhibitor, after stimulation by 10 μM arachidonic acid.

5. DISCUSSION

Although it has been reported that hsp27, p38, Akt and MK2 form a stable complex (Rane et al., 2001), how they interact with each other and how the interaction is regulated are not fully understood. The complex was identified by immunoprecipitation method using anti-Akt, anti-p38, anti-hsp27 and anti-MK2 antibodies. No FRET study on the complex has been performed. In particular, no one has observed the complex and the interaction within the complex in living cells. The present

FRET study provides new information about the complex. First, we observed a fluorescence resonance energy transfer from the hsp27-fused cyan fluorescent protein to the p38-fused yellow fluorescent protein in single living cells. When p38 is activated by H_2O_2-stimulation, such an energy transfer disappeared. This is consistent with the report that hsp27 dissociates from this complex when cell was stimulated. The question "Whether the dissociation of hsp27 from the complex is due to phosphorylation of hsp27 by p38-activated MK2 or phosphorylation of other components in this complex?" needs to be further investigated. Second, we observed that activation of Akt by activating PI3K signaling leads to loss of the interaction between hsp27 and p38. This could be the consequence of the phosphorylation of hsp27 by Akt on Ser-82 residue (Rane et al., 2003). However, we also observed that either SB203580, the inhibitor of p38, or wortmannin, the inhibitor of PI3K was able to prevent the interaction between p38 and hsp27 from being lost when cell was stimulated either by H_2O_2 or by arachidonic acid. The FRET experiment with these two inhibitors further proves that the activation of either p38 or Akt results in loss of the interaction of p38 with hsp27. This study may well demonstrate the regulatory role of Akt in the interaction of p38 with hsp27.

The study on the interaction of small hear shock protein 27 with MAP kinase p38 by FRET technology described in this chapter may provide a good example for showing the power of FRET technique. FRET imaging microscopy has been proved to be an extremely useful tool in the detection of protein-protein interactions and protein conformational changes in a single living cell. Future application of FRET imaging may involve cellular events coupled to specific molecular signaling processes, imaging in thick tissues or organisms using multiphoton excitation and detection of single-molecule FRET.

6. ACKNOWLEDGEMENTS

The work was supported by the National Natural Science foundation of China (No.60138010).

7. REFERENCES

Anderson, N. G., 1998, Co-immunoprecipitation, identification of interacting proteins, *Methods Mol.Biol.* 88: 35-45.
Bastiaens, P.I.H., and Squire, A., 1999, Fluorescence lifetime imaging microscopy: spatial resolution of biochemical processes in the cell, *Trends Cell Biol* .9:48-52.
Bastiaens, P.I.H., Pepperkok, R., 2000, Observing proteins in their natural habitat: the living cell, *Trends Biochem Sci.* 25: 631-637.
Bayerl ,C., Dorfner, B., Rzany, B., Fuhrmann, E., Coelho, C.C., Jung, E.G., 1999, Heat shock protein HSP 27 is expressed in all types of basal cell carcinoma in low and high risk UV exposure groups, *Eur.J.Dermatol.*, 9: 281-284.
Becker, W., Beergmann, A., Hink, M.A., Koenig, K., Benndo, K., and Biskup, C., 2004, Fluorescence lifetime imaging by time-correlated single-photon counting, *Microsc. Res.Tech.* 63: 58 – 66.
Burgering BM, Coffer PJ., 1995, Protein kinase B (c-Akt) in phosphatidylinositol-3-OH kinase signal transduction. *Nature.* 376: 599-602.
Carlsson K, Liljeborg A., 1998, Simultaneous confocal lifetime imaging of multiple fluorophores using the intensity-modulated multiplewavelength scanning (IMS) technique, *J Microsc.* 191: 119 – 127.
Chalfie, M., Tu, Y., Euskirchen, G., Ward, W.W., Prasher, D.C., 1994 Green fluorescent protein as a marker for gene expression, *Science* 263: 802-805.

Cole, M.J., Siegel, J., Dowling, R., Dayel, M.J., Parsons-Karavassilis, D., French, P.M., Lever, M.J., Sucharov, L.O., Neil, M.A., Juskaitis, R., Webb, S.E., Wilson, T., 2001, Time-domain whole-field fluorescence lifetime imaging with optical sectioning, *J Microsc.* 203: 246 - 257.

Ellenberg, J., Lippincott-Schwartz, J., Presley, J.F., 1999, Dual-colour imaging with GFP variants, *Trends Cell Biol.* 9: 52-56.

Fields, S., Song, O., 1989, A novel genetic system to detect protein-protein interactions, *Nature* 340:245-246.

Forster, T., 1948, Zwischenmoleculare Energiewanderung und Fluorescenz, *Ann. Phys.* 2: 57-75.

Freshney, N.W., Rawlinson, L., Guesdon, F., Jones, E., Cowley, S., Hsuan, J., and Saklatvala, J., 1994, Interlieukin-1 activates a novel protein kinase cascade that results in the phosphorylation of Hsp27, *Cell* 78: 1039-1049.

Hii, C.S.T., Moghadammi, N., Dunbar, A., and Ferrante A., 2001, Activation of the phosphatidylinositol 3-kinase-Akt/Protein kinase B signaling pathway in arachidonic acid-stimulated human myeloid and endothelial cells, *J.Biol.Chem.* 276: 27246-27255.

Khalid, H., Tsutsumi, K., Yamashita, H., Kishikawa, M., Yasunaga, A., Shibata, S., 1995, Expression of the small heat shock protein (hsp) 27 in human astrocytomas correlates with histologic grades and tumor growth fractions, *Cell Mol Neurobiol.* 15: 257-68.

Luban, J., and Goff, S. P., 1995, The yeast two-hybrid system for studying protein-protein interactions, *Curr.Opin.Biotech.* 6: 59-64.

Mitchell, A.C., Wall, J.E., Murray, J.G., Morgan, C.G., 2002, Direct modulation of the effective sensitivity of a CCD detector: A new approach to time-resolved fluorescence imaging, *J Microsc.* 206:225 - 232.

Mitseli, T., Spector, D.L., 1997, Application of the green fluorescent protein in cell biology and biotechnology, *Nat.Biotechnol.* 15: 961-964.

Miyawaki, A., Llopis, J., Heim, R. et al., 1997. Fluorescent indicators for Ca^{2+}-based on green fluorescent proteins and calmodulin, *Nature* 388: 882-885.

Nakatsue, T., Katoh, I., Nakamura, S., Takahashi, Y., Ikawa, Y., Yoshinaka, Y., 1998, Acute infection of Sindbis virus induces phosphorylation and intracellular translocation of small heat shock protein HSP27 and activation of p38 MAP kinase signaling pathway, *Biochem. Biophys. Res. Commun.* 253: 59-64.

Ng, T., Squire, A., Hansra, G. et al., 1999, Imaging protein kinase Ca activation in cells, *Science* 283: 2085-2089.

Pollok, B.A., Heim, R., 1999, Using GFP in FRET-based applications, Trends Cell Biol., 9, 57-60.

Rane, M.J., Coxon, P.Y., Powell, D.W., Webster, R., Klein, J.B., Pierce, W., Ping, P., McLeish, K.R., 2001, p38 Kinase-dependent MAPKAPK-2 activation functions as 3-phosphoinositide-dependent kinase-2 for Akt in human neutrophils, *J Biol Chem.* 276:3517-3523.

Rane, M.J., Pan, Y., Singh, S., Powell, D.W., Wu, R., Cummins, T., Chen, Q., McLeish, K. and Klein, J.B., 2003, Heat shock protein 27 controls apoptosis by regulating Akt activation, *J.Biol.Chem.* 278:27828-27835.

Ricci JE, Maulon L, Battaglione-Hofman V, Bertolotto C, Luciano F, Mari B, Hofman P, Auberger P., 2001, A Jurkat T cell variant resistant to death receptor-induced apoptosis, Correlation with heat shock protein (Hsp) 27 and 70 levels, *Eur Cytokine Netw.* 12:126-134.

van Roessel, P and Brand, A.H. 2001, Imaging into the future: visualizing gene expression and protein interactions with fluorescent proteins, *Nature Cell Biol.* 4:E15-E20.

Rouse, J., Cohen, P., Trigon, S., Morange, M., Alonso-Llamazares, A., Zamanillo, D., Hunt, T., and Nebreda, A., 1994 A novel kinase cascade trigged by stress and heat shock that stimulates MAPKAP kinase-2 and phosphorylation of the small heat shock proteins, *Cell* 78: 1027-1037.

Squire A, Verveer PJ, Bastiaens P.I.H., 2000, Multiple frequency fluorescence lifetime imaging microscopy, *J Microsc.* 197: 136 - 149.

Sytsma, J., Vroom, J.M., de Grauw, C.J., Gerritsen, H.C., 1998, Time-gated fluorescence lifetime imaging and microvolume spectroscopy using two-photon excitation, *J Microsc.* 191: 39 - 51.

Tsien, R.Y., 1998, The green fluorescent protein, *Ann. Rev.Biochem*, 67: 509-544.

Chapter 5

FUNCTIONAL OPTICAL COHERENCE TOMOGRAPHY: SIMULTANEOUS IN VIVO IMAGING OF TISSUE STRUCTURE AND PHYSIOLOGY

Zhongping Chen[1]

1. INTRODUCTION

Optical coherence tomography (OCT) is a recently developed imaging modality based on coherence-domain optical technology[1-4]. OCT takes advantage of the short coherence length of broadband light sources to perform micrometer-scale, cross-sectional imaging of biological tissue. OCT is analogous to ultrasound imaging except that it uses light rather than sound. The high spatial resolution of OCT enables noninvasive *in vivo* "optical biopsy" and provides immediate and localized diagnostic information. OCT was first used clinically in ophthalmology for the imaging and diagnosis of retinal disease[2]. Recently, it has been applied to imaging subsurface structure in skin, vessels, oral cavities, as well as respiratory, urogenital, and GI tracts[2].

OCT uses coherent gating of backscattered light for tomographic imaging of tissue structure. Variations in tissue scattering due to inhomogeneities in the optical index of refraction provide imaging contrast. However, in many instances and especially during early stages of disease, the change in tissue scattering properties between normal and diseased tissue is small and difficult to measure. One of the great challenges for extending clinical applications of OCT is to find more contrast mechanisms that can provide physiological information in addition to morphological structure. A number of extensions of OCT capabilities for functional imaging of tissue physiology have been developed. Doppler OCT, also named optical Doppler tomography (ODT), combines the Doppler principle with OCT to obtain high-resolution tomographic images of tissue structure and blood flow simultaneously[5-13]. Spectroscopic OCT combines spectroscopic analysis with OCT to obtain the depth resolved tissue absorption spectra[14-16]. Polarization sensitive OCT (PS-OCT) combines polarization sensitive detection with

---

[1] Biomedical Engineering Beckman Laser Institute University of California, Irvine Irvine, CA 92612
E-mail address: zchen@bli.uci.edu.

OCT to determine tissue birefringence[17-21]. Second harmonic optical coherence tomography combines second harmonic generation with coherence gating to obtain images with molecular contrast[22]. These functional extensions of OCT provide clinically important information on tissue physiology, such as tissue blood perfusion, oxygen saturation, hemodynamics, and structural remodeling. Each provides several potential clinical applications, such as vasoactive drug screening, tissue viability and burn depth determination, tumor angiogenesis studies and tumor diagnosis, bleeding ulcer management, and ocular pathology evaluation[10,11,20,23-25]. This chapter reviews the principle of OCT and Functional OCT (F-OCT) and highlights some of the results obtained in the OCT Laboratory at the Beckman Laser Institute.

2. OPTICAL COHERENCE TOMOGRAPHY

One of the challenges for tomographic imaging of tissue using visible or near-infrared light is tissue optical scattering. When photons propagate inside tissue, they undergo multiple scattering. The mean free path of photons in skin, for example, is 5 μm. If one focuses the light 1 mm below the skin surface, most photons undergo multiple scattering and never reach the target point. Less than one in a million photons will reach the target, undergo a single backscattering, and return to the detector. Most photons that are backscattered and reach the detector are multiple scattered photons that do not carry information on the target point. Therefore, the challenge for tomographic imaging of tissues is to separate the singly backscattered photons that carry information about the target point from multiple scattered photons that form background noise.

OCT uses coherent gating to discriminate single scattered photons from multiple scattered photons. OCT is based on a Michelson interferometer with a broadband light source (Figure 1). Light from a broadband partial coherent source is incident on the beam splitter and split equally between reference and target arms of the interferometer. Light backscattered from the turbid sample recombines with light reflected from the reference arms and forms interference fringes. High axial spatial resolution is possible because interference fringes are observed only when the pathlength differences between the sample arm and reference arm are within the coherence length of the source. Axial scans

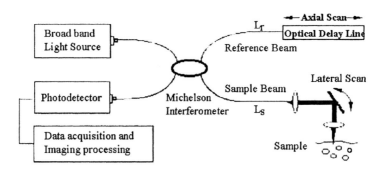

Figure 1. Schematic of OCT system consists of a fiber based Michelson interferometer with a partially coherent light source.

are performed by scanning the reference arms, and lateral scans are performed by scanning the sample beam. A two-dimensional, cross-sectional image is formed by performing the axial scan followed by a lateral scan. Axial resolution is determined by coherence length of the source, and lateral resolution is determined by the numerical aperture of the sampling focus lens. To illustrate the coherence gating of the broadband light source, we consider the interference fringe generated in the Michelson interferometer with a broadband light source. Let us denote $U(t)$ as a complex-valued analytic signal of a stochastic process representing the field amplitude emitted by a low coherent light source, and $\overline{U}(v)$ as the corresponding spectral amplitude at optical frequency v. The amplitude of a partially coherent source light coupled into the interferometer at time t is written as a harmonic superposition

$$U(t) = \int_0^\infty \overline{U}(v)e^{2\pi i v t} dv. \tag{1}$$

Because the stochastic process of a partially coherent light source is stationary, the cross spectral density of $\overline{U}(v)$ satisfies

$$\left\langle \overline{U}^*(v)\overline{U}(v') \right\rangle = S_o(v)\delta(v-v'), \tag{2}$$

where $S_o(v)$ is the source power spectral density, and $\delta(v-v')$ is the Dirac delta function. Assuming that light couples equally into the reference arm and sample arm with spectral amplitude of $\overline{U}_o(v)$, the light coupled back to the detector from the reference, $\overline{U}_r(v)$, and sample, $\overline{U}_s(v)$, is given by Eqs. (3) and (4), respectively:

$$\overline{U}_r(v) = e^{2\pi i v(2L_r + L_d)/c} K_r(v)e^{i\alpha_r} \overline{U}_o(v), \tag{3}$$

$$\overline{U}_s(v) = e^{2\pi i v(2L_s + L_d)/c} K_s(v)e^{i\alpha_s} \overline{U}_o(v), \tag{4}$$

where L_r and L_s are the optical pathlengths from the beam splitter to the reference mirror and sample, respectively, L_d is the optical pathlength from the beam splitter to the detector, and $K_r(v)e^{i\alpha_r(v)}$ and $K_s(v)e^{i\alpha_s(v)}$ are the amplitude reflection coefficients of light backscattered from the reference mirror and turbid sample, respectively.

The total power detected at the interferometer output is given by a time-average of the squared light amplitude

$$P_d(\tau) = \left\langle \left| U_s(t) + U_r(t+\tau) \right|^2 \right\rangle, \tag{5}$$

where τ is the time delay between light traveled in the sample and reference arms. Combining harmonic expansions for $U_S(t)$ and $U_r(t)$ and applying Eq. (2) when computing a time-average, total power detected is a sum of three terms representing reference (I_r), sample (I_s), and the interference fringe intensity ($\Gamma_{OCT}(\tau)$),

$$P_d(\tau) = \int_0^\infty \left(P_r(v) + P_s(v) + P_{OCT}(v) \right) dv = I_r + I_s + \Gamma_{OCT}(\tau), \tag{6}$$

with

$$P_r(v) = S_o(v)|K_r(v)|^2, \tag{7}$$

$$P_s(v) = S_o(v)|K_s(v)|^2, \tag{8}$$

$$P_{OCT}(v) = 2S_o(v)K_r(v)K_s(v)\cos[2\pi v(\Delta_d/c + \tau) + \alpha_s(v) - \alpha_r(v)], \tag{9}$$

and

$$I_r = \int_0^\infty P_r(v)dv, \tag{10}$$

$$I_s = \int_0^\infty P_s(v)dv, \tag{11}$$

$$\Gamma_{OCT}(\tau) = \int_0^\infty P_{OCT}(v)dv, \tag{12}$$

where Δ_d is the optical pathlength differences between light traveled in sample and reference arms. To simplify the computation, we assume α_s and α_r are constants over the source spectrum and can be neglected. The spectral domain fringe signal, $P_{OCT}(v)$, is simplified to:

$$P_{OCT}(v) = 2S_o(v)K(v)_r K_s(v)\cos[2\pi v(\Delta_d/c + \tau)]. \tag{13}$$

The corresponding time domain signal, $\Gamma_{OCT}(\tau)$, is given by:

$$\Gamma_{OCT}(\tau) = 2\int_0^\infty S_o(v)K(v)_r K_s(v)\cos[2\pi v(\Delta_d/c + \tau)]dv. \tag{14}$$

A comparison of Eqs. (13) and (14) shows that there is a Fourier transformation relation between spectral domain and time domain signals. Consequently, there are two methods to acquire the OCT signal: the spectral domain method and the time domain method.

In the spectral domain method, the reference mirror is fixed, and there is no depth scan ($\tau=0$). The spectral domain fringe signal, $P_{OCT}(v)$, is obtained either by a spectrometer at the detection arm, or by a frequency sweeping light source[26-28]. The depth resolved fringe signal, $\Gamma_{OCT}(\tau)$, is determined from the spectral domain signal by a Fourier transformation. The spectral domain technique has the advantage of high signal to noise ratio in comparison to the time domain method. In addition, there is no need to scan the reference arm[2,29]. However, it requires a high-speed spectrometer or a high-speed spectral sweeping source.

In the time domain method, a delay line is incorporated in the reference arm to generate delay. In the case where K_s and K_r are spectrally independent over the source spectrum, the interference fringe term, $\Gamma_{OCT}(\tau)$, is simplified to (assuming $\Delta_d=0$ initially):

$$\Gamma_{OCT}(\tau) = 2K_r K_s \int_0^\infty S_o(v)\cos[2\pi v\tau]dv. \tag{15}$$

Furthermore, if the light source has a Gaussian power spectrum with a full width at half maximum (FWHM) Δv:

$$S_o(v) \propto \exp\left[-4\ln 2\frac{(v-v_0)^2}{\Delta v^2}\right], \tag{16}$$

the interference fringe term reduces to:

$$\Gamma_{OCT}(\tau) \propto \exp\left[-4\ln 2 \frac{\tau^2}{\tau_c^2}\right]\cos[2\pi\nu_0\tau], \quad (17)$$

where τ_c is the coherence time of the light source given by:

$$\tau_c = \frac{4\ln 2}{\pi\Delta\nu}. \quad (18)$$

Because OCT detects backscattered light, the round trip coherence length is related to the coherence time by:

$$L_c = \frac{\tau_c}{2}c = \frac{2\ln 2}{\pi}\frac{\lambda_0^2}{\Delta\lambda}, \quad (19)$$

where c is the speed of light, λ_0 is the center wavelength of the broadband source, and $\Delta\lambda$ is the FWHM of the source spectra. Using coherence length, the interferometer fringe term can be written as:

$$\Gamma_{OCT}(\Delta L) \propto \exp\left[-4\ln 2 \frac{\Delta L^2}{L_c^2}\right]\cos\left[4\pi\frac{\Delta L}{\lambda_0}\right]. \quad (20)$$

Γ_{OCT} has an interference fringe term with a Gaussian envelope. Figure 2 shows a normalized plot of the interference fringe term using a partial coherent source with coherence lengths of 1 μm and 10 μm, respectively. The figure shows that when a partially coherent light source is used, the fringe detected has a maximum amplitude when the pathlengths in the reference and sample arms are matched ($\Delta L=0$). The amplitude decays when the pathlength difference increases. The signal decays to half of its maximum amplitude when the pathlength difference is equal to the coherence length of the source. The fringe signal decays to a negligible level when the pathlength difference between the reference and sample arms increases to several times of coherence length of the source. Therefore, OCT is based on the principle of coherence gating. One

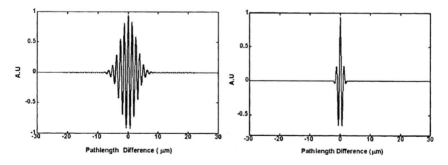

Figure 2. Simulated interference fringe signals from Michelson interferometer with partial coherence light sources at 1.3 μm and coherence lengths of 10 μm (left) and 1 μm (right), respectively.

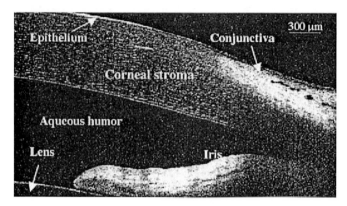

Figure 3. High-resolution ophthalmic image of the anterior chamber of a rabbit eye.

can observe the interference fringe only when the light pathlengths in the reference and smple arms are matched within the coherence length of the source. By scanning the reference arm and measuring the fringe, one can map the location of the scattering particle in a scattering sample.

The axial resolution of OCT is determined by the coherent length of the source. Most OCT systems currently used in clinical trials have an axial resolution of 10 μm, which is not good enough to resolve subcelluar structure. High resolution OCT with axial resolution on the order of 1 μm has been demonstrated by several groups using ultra-broadband light sources[30]. Figure 3 shows an image of the anterior chamber of a rabbit eye[4,31]. The cornea, iris, and conjunctiva can be clearly visualized. The entire image depth is 2 mm with an axial resolution of 1.3 μm. The epithelium layer can be clearly delineated even though the thickness is only 15 μm.

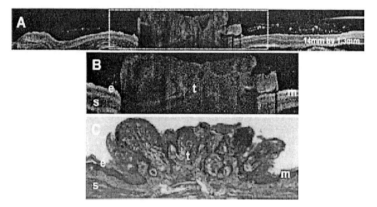

Figure 4. OCT image of hamster cheek pouch containing squamous cell carcinoma tumor: e-squamous epithelum, m-mucosa, s-submucosa, t- fungiform malignant tissue. Corresponding histology is shown in (C).

There are many potential clinical applications of OCT. We have imaged a number of organ sites in both animal and human subjects, including skin, brain, cochlea, retina, and mucosa in the respiratory and GI tracts. Figure 4 illustrates the ability of OCT to detect malignancies. We have obtained in vivo OCT image of the squamous cell carcinoma in a hamster cheek pouch (Figure 4). In the normal region, squamous epithelium, mucosa, submucosa, and basement membrane can be clearly identified. The tumor region shows loss of normal epithelial stratification with breakdown of the basement membrane and epithelial invasion into subepithelial layers. OCT images containing cancers appear to have more scattering and absorptive regions, compared to normal tissue, reducing overall depth of penetration of images. For comparison, we also include the histology image from the same region. The epithelium is folded, as are the mucosa and basement membrane. The basement membrane is no longer intact due to the invasion of malignant cells into the connective tissue below. The feature from the OCT image corresponds well with the histology image.

3. FUNCTIONAL OPTICAL COHERENCE TOMOGRAPHY

The contrast mechanism for OCT structure images is based on variations in tissue scattering due to inhomogeneities in the optical index of refraction. However, in many instances and especially in the early stages of disease, the change in tissue scattering properties between normal and diseased tissue is small. F-OCT extends OCT capability by using additional contrast mechanisms. F-OCT can provide clinically important physiological information in addition to morphological structural image. Doppler OCT, PS-OCT, and Second harmonic OCT will be briefly discussed in this section.

3.1 Doppler OCT

Doppler OCT, also named optical Doppler tomography (ODT), combines the Doppler principle with OCT to obtain high resolution tomographic images of static and moving constituents in highly scattering biological tissues[10,12,20,32]. When light backscattered from a moving particle interferes with the reference beam, a Doppler frequency shift (f_D) occurs in the interference fringe:

$$f_D = \frac{1}{2\pi}(\mathbf{k}_s - \mathbf{k}_i) \bullet \mathbf{v}$$

(21)

where $\mathbf{k}_i$ and $\mathbf{k}_s$ are wave vectors of incoming and scattered light, respectively, and $\mathbf{v}$ is the velocity vector of the moving particle. Since ODT measures the backscattered light, assuming the angle between flow and sampling beam is θ, the Doppler shift equation is simplified to:

$$f_D = \frac{2V\cos\theta}{\lambda_0}, \tag{22}$$

where λ_0 is the vacuum center wavelength of the light source. Longitudinal flow velocity (velocity parallel to the probing beam) can be determined at discrete, user-specified locations in a turbid sample by measurement of the Doppler shift[6,7,12,32]. Transverse flow velocity can also be determined from the broadening of the spectral bandwidth due to the finite numeric aperture of the probing beam[13].

The first use of coherence gating to measure localized flow velocity was reported in 1991, where the one-dimensional velocity profile of the flow of particles in a duct was measured[33]. In 1997, the first two-dimensional *in vivo* ODT imaging was reported using the spectrogram method[6,7]. The spectrogram method uses a short time fast Fourier transformation (STFFT) or wavelet transformation to determine the power spectrum of the measured fringe signal[5-8,10,11]. Although spectrogram methods allow simultaneous imaging of *in vivo* tissue structure and flow velocity, the velocity sensitivity is limited for high speed imaging. Phase-resolved ODT was developed to overcome these limitations[12,32]. This method uses the phase change between sequential A-line scans for velocity image reconstruction[12,32,34,35]. Phase-resolved ODT decouples spatial resolution and velocity sensitivity in flow images and increases imaging speed by more than two orders of magnitude without compromising spatial resolution and velocity sensitivity[12,32]. The minimum flow velocity that can be detected using an A-line scanning speed of 1000 Hz is as low as 10 μm/s while maintaining a spatial resolution of 10 μm.

When there is a moving particle, light scattered from a moving particle is equivalent to a moving phase front; therefore, Δ_d in Eq. (13) can be written as:

$$\Delta_d = \Delta + 2\bar{n}v_z t , \qquad (23)$$

where Δ is the optical pathlength difference between light in the sampling and reference arms, v_z is the velocity of a moving particle parallel to the probe beam, $\bar{n}$ is the refractive index of flow media.

Equations (13, 14) that describe the frequency and time domain OCT signals have to be modified to reflect the moving phase front due to moving scattering particles:

$$P_{ODT}(v) = 2S_o(v)K_r(v)K_s(v)\cos[2\pi v((\Delta + 2\bar{n}v_z t)/c + \tau)] . \qquad (24)$$

$$\Gamma_{ODT}(\tau) = 2\int_0^\infty S_o(v)K_r(v)K_s(v)\cos[2\pi v((\Delta + 2\bar{n}v_z t)/c + \tau)]dv . \qquad (25)$$

Equation 25 shows that moving particles induce a phase change in the fringe signal and the Doppler frequency shift can be determined from the phase change between sequential scans.

The phase information of the fringe signal can be determined from the complex analytical signal $\dot{\Gamma}(t)$, which is determined through analytic continuation of the measured interference fringe function, $\Gamma(t)$, using a Hilbert transformation[12,35]:

$$\dot{\Gamma}(t) = \Gamma(t) + \frac{i}{\pi} P \int_{-\infty}^{\infty} \frac{\Gamma(\tau)}{\tau - t} d\tau = A(t)e^{i\phi(t)} \qquad (26)$$

where P denotes the Cauchy principle value, i is the complex number, and $A(t)$ and $\phi(t)$ are the amplitude and phase of $\dot{\Gamma}(t)$, respectively. Because the interference signal $\Gamma(t)$ is quasi-monochromatic, the complex analytical signal can be determined by[35]:

$$\dot{\Gamma}(t) = 2\int_0^{+\infty+\tau}\int_0^{\tau}\Gamma(t)\exp(-2\pi i v t)dt\,\exp(2\pi i v t)dv\,,\tag{27}$$

where τ is the time duration of the fringe signal in each axial scan.

The Doppler frequency shift (f_n) at nth pixel in the axial direction is determined from the average phase shift between sequential A-scans. This can be accomplished by calculating the phase change of sequential scans from the individual analytical fringe signal[12,32]:

$$f_n = \frac{\Delta\phi}{2\pi T} = \frac{1}{2\pi T}\sum_{m=(n-1)M}^{nM}\sum_{j=1}^{N}\left[\tan^{-1}\left(\frac{\operatorname{Im}\dot{\Gamma}_{j+1}(t_m)}{\operatorname{Re}\dot{\Gamma}_{j+1}(t_m)}\right) - \tan^{-1}\left(\frac{\operatorname{Im}\dot{\Gamma}_j{}^*(t_m)}{\operatorname{Re}\dot{\Gamma}_j{}^*(t_m)}\right)\right]\tag{28}$$

Alternatively, the phase change can also be calculated by the cross-correlation method [12, 32]:

$$f_n = \frac{1}{2\pi T}\tan^{-1}\left(\frac{\operatorname{Im}\left[\sum_{m=(n-1)M}^{nM}\sum_{j=1}^{N}\dot{\Gamma}_j(t_m)\cdot\dot{\Gamma}_{j+1}(t_m)\right]}{\operatorname{Re}\left[\sum_{m=(n-1)M}^{nM}\sum_{j=1}^{N}\dot{\Gamma}_j(t_m)\cdot\dot{\Gamma}_{j+1}(t_m)\right]}\right),\tag{29}$$

where $\dot{\Gamma}_j(t_m)$ and $\dot{\Gamma}_j{}^*(t_m)$ are the complex signal at axial time t_m corresponding to the *j*th A-scan and its respective conjugate; $\dot{\Gamma}_{j+1}(t_m)$ and $\dot{\Gamma}_{j+1}{}^*(t_m)$ are the complex signal at axial time t_m corresponding to the next A-scan and its respective conjugate; M is an even number that denotes the window size in the axial direction for each pixel; N is the number of sequential scans used to calculate the cross correlation; and T is the time duration between A-scans. Because T is much longer than the pixel time window within each scan used in spectrogram method, high velocity sensitivity can be achieved.

The transverse flow velocity can be determined from the standard deviation of the Doppler fringe signal:

$$\sigma^2 = \frac{1}{2\pi^2 T^2}\left(1 - \frac{\left|\sum_{m=(n-1)M}^{nM}\sum_{j=1}^{N}\dot{\Gamma}_j(t_m)\cdot\dot{\Gamma}_{j+1}{}^*(t_m)\right|}{\frac{1}{2}\sum_{m=(n-1)M}^{nM}\sum_{j=1}^{N}[\dot{\Gamma}_j(t_m)\cdot\dot{\Gamma}_j{}^*(t_m) + \dot{\Gamma}_{j+1}(t_m)\cdot\dot{\Gamma}_{j+1}{}^*(t_m)]}\right).\tag{30}$$

Equation (24) shows that the spectral domain Doppler signal, $P_{ODT}(v)$, also contains information on both locations and velocity of moving particles. A comparison of Eqs. (24) and (25) shows that there is a Fourier transformation relation between spectral domain and time domain signals. The phase shift due to the moving particle can be determined from the Fourier transformations of two sequential spectral domain fringe signals. The Doppler frequency can then be calculated using Eqs. (28) or (29) of the phase-resolved method.

Figure 5 shows the first *in vivo* structural and blood flow images from a rodent skin[6]. Cross-sectional structural (Fig. 5A) and velocity images (Figs. 5B and C) are obtained simultaneously. The presence of vessel-like circular features can be observed in the structural image (Fig. 5A). Velocity images of blood flow moving in opposite directions, as determined by the sign of the Doppler frequency shift, are shown in Figures 5B and C. Blood flow in two small veins (Fig. 5B) and an artery (Fig. 5C) is clearly identified.

Due to its exceptionally high spatial resolution and velocity sensitivity, Doppler OCT has a number of applications in biomedical research and clinical medicine. Several clinical applications of Doppler OCT have been demonstrated in our laboratory, including screening vasoactive drugs, monitoring changes in image tissue morphology and hemodynamics following pharmacological intervention and photodynamic therapy, evaluating the efficacy of laser treatment in port wine stain patients, assessing the depth of burn wounds, and mapping cortical hemodynamics for brain research[10-12,23,32,34]. Application of Doppler OCT in ophthalmology was demonstrated[36], and recently, endoscopic application of Doppler OCT for imaging blood flow in the gastrointestinal tracts was also reported[37]. Figure 6 shows Doppler OCT structural and flow velocity images of a patient with PWS before and after laser treatment, respectively. For comparison, we also include a histology picture taken at the same site. The vessel location from the Doppler OCT measurement and histology agree very well.

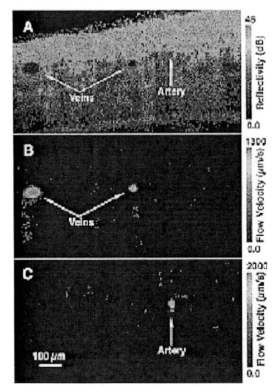

Figure 5. Doppler OCT images of *in vivo* blood flow in rodent skin: A, structural image. B, velocity image of venous blood flow (into the page); C, velocity image of arterial blood flow (out of the page), (from Ref. [6]).

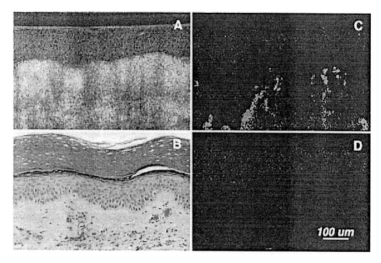

Figure 6. Doppler OCT images taken *in situ* from PWS human skin. A: structural image; B: histological section; C: image before laser treatment; and D: image after laser treatment (from Ref. [23]).

Furthermore, the destruction of the vessel by laser can be identified since no flow appears on the Doppler flow image after laser treatment. This result indicates that Doppler OCT can provide a fast, semi-quantitative evaluation of the efficacy of PWS laser therapy *in situ* and in real-time.

3.2 Polarization Sensitive OCT

PS-OCT combines polarization sensitive detection with OCT to determine tissue birefringence. PS-OCT can obtain enhanced image contrast and additional physiological information by studying polarization properties of biological tissues. Many biological tissues, such as tendon, muscle, nerve, bone, cartilage, and skin, exhibit birefringence properties. Consequently, PS-OCT can provide novel contrast mechanisms that are directly related to the physiological conditions of these tissues. The first PS-OCT using a free-space Michelson interferometer was reported in 1992[17]. A fiber based PS-OCT system was demonstrated in 2000[20]. Since 1992, a number of potential clinical applications of PS-OCT have been reported[17,18,20,21,24,38,39].

The polarization state of a light beam can be described by a Stokes vector S as:

$$S = \begin{bmatrix} I \\ Q \\ U \\ V \end{bmatrix} = \begin{bmatrix} a_x^2 + a_y^2 \\ a_x^2 - a_y^2 \\ 2a_x a_y \cos\varphi \\ 2a_x a_y \sin\varphi \end{bmatrix}, \tag{31}$$

where I, Q, U, V are Stokes parameters, a_x and a_y are amplitudes of two orthogonal components of the electric vector and φ represents the phase difference between the two components.

The effect of an optical device on the polarization of light can be characterized by a 4×4 Mueller matrix. The matrix acts on the input state S_1 to give the output state S_2:

$$S_2 = \begin{vmatrix} I_2 \\ Q_2 \\ U_2 \\ V_2 \end{vmatrix} = MS_1 = \begin{vmatrix} M_{00} & M_{01} & M_{02} & M_{03} \\ M_{10} & M_{11} & M_{12} & M_{13} \\ M_{20} & M_{21} & M_{22} & M_{23} \\ M_{30} & M_{31} & M_{32} & M_{33} \end{vmatrix} \begin{vmatrix} I_1 \\ Q_1 \\ U_1 \\ V_1 \end{vmatrix}. \tag{32}$$

In the Poincaré sphere representation, a polarization state S can be represented by a point (Q,U,V). The Poincaré sphere representation provides a convenient method to evaluate changes of Stokes vectors. Determination of either the Stokes vector or the Mueller matrix allows full quantification of birefringence properties of the biological sample. PS-OCT imaging of the Stokes vector and Muller matrix has been demonstrated[18,20,21,39].

The fiber-based, high-speed PS-OCT system is shown in Figure 7[19]. A 1310 nm partially coherent light source with a FWHM bandwidth of 80 nm was used. As the source in the fiber-based PS-OCT system is unpolarized, light delivered to the sample thus has components of any polarization states. Due to the coherence detection intrinsic to OCT, backscattered light that originates from incident light in different polarization states can be determined by controlling polarization states in the reference arm. Modulating the reference light in four orthogonal polarization states allows coherent detection of backscattered light from the sample under equivalent illumination in four different polarization states. Light entering a 2x2 fiber splitter is divided equally into the sample and reference arms of the Michelson interferometer. In the reference arm, a rapid scanning optical delay (RSOD) line is aligned such that only group delay scanning at 500 Hz is generated without phase modulation. A stable ramp phase modulation at 500 kHz is generated using an E-O modulator for heterodyne detection. When light returned from the RSOD passes through the phase modulator, it polarizes light to 45° with respect to the optical axis of the crystal in the polarization modulator. A probe with a collimator and an infinity-corrected objective driven by a translation stage is employed in the sample arm. The fringe signals from the two polarization channels are detected by two photo-detectors, then high pass filtered, amplified and digitized by a 12-bit analogue to digital conversion board. A polarization modulator is used to control the polarization state of light in the reference arm, which rapidly varies between states orthogonal in the Poincaré sphere at 500 Hz. The choice of orthogonal polarization states in the Poincaré sphere is important because it ensures that the birefringence measurements will be independent of the orientation of the optical axis in the sample. For every polarization state controlled by the polarization modulator, the A-scan signals corresponding to the two orthogonal polarization diversity channels are digitized. Considering the quasi-monochromatic light beam, the coherence matrix can be calculated from the complex electrical field vector from these two channels:

$$J = \begin{vmatrix} \langle E_H^*(t)E_H(t) \rangle & \langle E_H^*(t)E_V(t) \rangle \\ \langle E_V^*(t)E_H(t) \rangle & \langle E_V^*(t)E_V(t) \rangle \end{vmatrix} = \begin{bmatrix} J_{HH} & J_{HV} \\ J_{VH} & J_{VV} \end{bmatrix}, \tag{33}$$

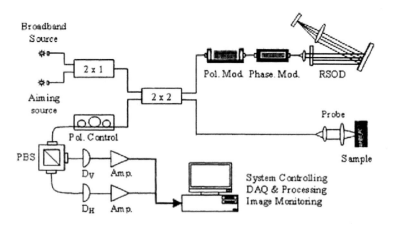

Figure 7. Fiber-based PS-OCT system.

where E_H, E_V are the components of the complex electric field vector corresponding to horizontal and vertical polarization channels, respectively, and E_H^*, E_V^* are their conjugates, respectively. The Stokes vector can be derived from the coherence matrix[20]:

$$S_0 = J_{HH} + J_{VV} \qquad S_2 = S_0^{-1}(J_{HV} + J_{VH})$$
$$S_1 = S_0^{-1}(J_{HH} - J_{VV}) \qquad S_3 = iS_0^{-1}(J_{VH} - J_{HV}) \qquad (34)$$

where S_0, S_1, S_2, and S_3 are the four components of the Stokes vector. S_1, S_2, and S_3 are the normalized coordinates of the Stokes vector in the Poincaré sphere characterizing the polarization state of the backscattered light, and S_0 is the module of the Stokes vector characterizing light intensity. The Stokes vector corresponding to each of the input polarization state for the nth pixel in the jth A-can can be calculated as:

$$S_{0,n} = \sum_{m=(n-1)M}^{nM} \left[\tilde{F}_j^H(t_m)\tilde{F}_j^{H*}(t_m) + \tilde{F}_j^V(t_m)\tilde{F}_j^{V*}(t_m)\right]$$
$$S_{1,n} = S_{0,n}^{-1} \sum_{m=(n-1)M}^{nM} \left[\tilde{F}_j^H(t_m)\tilde{F}_j^{H*}(t_m) - \tilde{F}_j^V(t_m)\tilde{F}_j^{V*}(t_m)\right]$$
$$S_{2,n} = S_{0,n}^{-1} \sum_{m=(n-1)M}^{nM} 2\text{Re}\left(\tilde{F}_j^{H*}(t_m)\tilde{F}_j^V(t_m)\right) \qquad (35)$$
$$S_{3,n} = S_{0,n}^{-1} \sum_{m=(n-1)M}^{nM} 2\text{Im}\left(\tilde{F}_j^{H*}(t_m)\tilde{F}_j^V(t_m)\right)$$

where M is an integer number that determines the size of the average in the axial direction for each pixel, $\tilde{F}_j^H(t_m)$ and $\tilde{F}_j^V(t_m)$ are complex signals detected from the two orthogonal polarization channels at axial time t_m, for the jth A-scan, $\tilde{F}_j^{H*}(t_m)$ and $\tilde{F}_j^{V*}(t_m)$ are their conjugates, respectively.

In order to measure the birefringence properties of the sample accurately, four states of light polarization are generated for each lateral pixel. For each polarization state, one A-line scan is performed. Therefore, a total of four A-line scans are used to calculate the Stokes vectors, polarization diversity intensity and birefringence images simultaneously. For a sample with an assumed linear birefringence, there exists two eigenwaves that are polarized along the projected fast and slow axes of the sample normal to the propagation direction of incident light. The Stokes vectors of these eigenwaves determine a rotation axis in the equator plane of a Poincaré sphere. The effect of birefringence is to rotate the Stokes vector about this axis through an angle that is equal to the phase retardation of the sample. Conversely, the rotation axis can be determined from the known polarization states of incident and backscattered light at the sample location. From the Stokes vectors for the four states of light polarization, the structural image is obtained by averaging the four S_0. The phase retardation image, which characterizes the accumulated birefringence distribution in the sample, is calculated by the rotation of the Stokes vectors in the Poincaré sphere.

The simultaneous Stokes vectors and phase retardation images of a dog ligament are shown in Figure 8. Stokes vectors corresponding to four different polarization states are shown in the figure on the left. From Stokes vectors image, phase retardation image can be calculated as shown in the figure on the right. The banded structure in the phase retardation image indicates the accumulated tissue birefringence distribution in the rat-tail tendon. In addition, differential optical axis and birefringence images can also be calculated[40].

There are a number of clinical applications for PS-OCT, including burn depth determination and retinal fiber nerve layer thinckness evaluation. Preliminary results indicate clear correlation between the skin birefringence and burn depth. Figure 9 shows *in vivo* structural and phase retardation images from a rat skin burned at 100 °C for 10 seconds. Although the structure image shows very little contrast, the phase retardation image clearly shows the contrast between the normal and burned regions. As light propagates into the tissue, the retardation increases due to tissue birefringence. For normal tissue, it only takes less than 20 µm thick tissue to cause a phase retardation of 90^0. However, for the burned tissue, it takes much thicker tissue to cause the same phase retardation. The decrease of birefringence in thermally damaged tissue can clearly be seen in PS-OCT scans when compared to normal rat skin.

3.3 Second Harmonic OCT

Second harmonic OCT (SH-OCT) combines second harmonic generation with coherence gating for high resolution tomographic imaging of tissue structures with molecular contrast. Optical second harmonic generation (SHG) is the lowest order nonlinear optical process where the second order nonlinear optical susceptibility is responsible for the generation of light at second harmonic frequency. Because second order nonlinear optical susceptibility is very sensitive to electronic configurations, molecular structure and symmetry, local morphology, and ultrastructures, SHG provides molecular contrast for the coherence image.

We have demonstrated, to the best of our knowledge, the first SH-OCT image[22]. The schematic diagram for SH-OCT is shown in the left panel of Figure 10. The light source was a Kerr-lens mode-locked Ti:sapphire laser pumping a nonlinear fiber to broaden the spectrum. A polarizing beam splitter (PBS) split the input beam into two

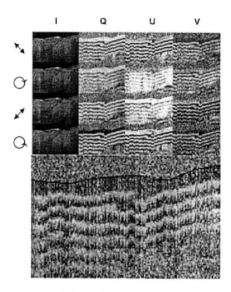

Figure 8. Stokes vectors images (left) and phase retardation image (right) in fresh ligament.

arms of the interferometer. In the reference arm, we used a 0.1 mm thick nonlinear crystal BBO oriented for type I phase matching to convert the input radiation to second harmonics at 400 nm. Both SH and fundamental waves were then reflected by a metal mirror (M2) mounted on a motorized translation stage, which acted as the delay line in this OCT system. The back-reflected radiation was partially reflected by a broadband non-polarizing beam splitter (BS1) and directed into the broadband, non-polarizing combining beam splitter (BS2). In the signal arm after the PBS, the fundamental radiation passed a half-wave plate (HWP3) and was focused onto the sample by a low numerical aperture lens (L1, N.A.=0.2, f=31.8mm). When the sample had second order nonlinear properties, the fundamental radiation generated second harmonics. Back-reflected second harmonic and fundamental waves were collimated by the same

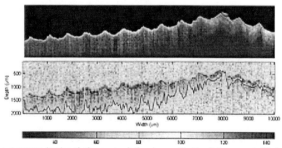

Figure 9. *In vivo* structure (top) and phase retardation image (bottom) from a rat skin burned with at 100 °C for 10 seconds. The black contour line in the phase retardation image demarcates the depth at which 90° phase retardation has been reached with respect to the incident polarization (image size 2 mm x 10 mm).

lens and directed by a dichroic beam splitter (DBS) toward BS2. The dichroic beam splitter reflected maximum second harmonic radiation and about 5% of the fundamental radiation. The radiation from signal and reference arms was recombined after passing BS2. By changing the optical path delay in the reference arm, the pulses overlapped temporally and interference fringes at fundamental and second harmonic wavelengths were generated. The harmonic interference fringe signal was detected by a photomultiplier tube (PMT) after passing through a short-pass filter (F2) and a 400 nm band-pass filter (F3). The fundamental interference fringe signal was detected by a photodiode (PD) after passing through a long-pass filter (F4). We also inserted a pair

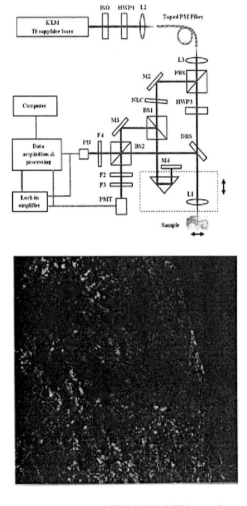

Figure 10. Schematic of experiment set-up for SH-OCT (left); SH-OCT image of a rat tendon (right). Image size 250 x250 μ m².

of prisms made from fused silica into the signal arm to compensate for the group-velocity dispersion of the fundamental and harmonic waves, thus enabling simultaneous observation of SH-OCT and conventional OCT signals.

SH-OCT image of a rat tendon is shown in the right panel of Figure 10. Compared with conventional OCT performed at fundamental wavelengths, SH-OCT offers enhanced molecular contrast and spatial resolution. It is also an improvement over existing second harmonic scanning microscopy technology as the intrinsic coherence gating mechanism enables the detection and discrimination of SH signals generated at deeper locations. The enhanced molecular contrast of SH-OCT extends conventional OCT's capability for detecting small changes in molecular structure. SH-OCT is promising for the diagnosis of cancers and other diseases at an early stage when changes in tissue and molecular structure are small.

4. CONCLUSIONS

OCT is a rapidly developing imaging technology with many potential applications. Through different contrast enhancement mechanisms, F-OCTprovides clinically important physiological information that is not available in the structure image. Integration of F-OCT, such as Doppler OCT, PS-OCT, spectroscopic OCT, and SH-OCT, can greatly enhance potential applications of this technology. Given the noninvasive nature and exceptionally high spatial resolution and velocity sensitivity, functional OCT that can simultaneously provide tissue structure, blood perfusion, birefringence, and other physiological information has great potential for basic biomedical research and clinical medicine.

5. ACKNOWLEDGMENTS

I would like to thank many of my colleagues who have contributed to the functional OCT project at the Beckman Laser Institute and Center for Biomedical Engineering at UCI, particularly to the students and postdoctoral fellows. I also want to acknowledge the research grants awarded from the National Institutes of Health (EB-00293, NCI-91717, RR-01192 and EB-00255), the National Science Foundation (BES-86924), the Whitaker Foundation (WF-23281) and the Defense Advanced Research Program Agency (Bioflip program). Institutional support from the Air Force Office of Scientific Research (F49620-00-1-0371), and the Beckman Laser Institute Endowment are also gratefully acknowledged.

6. REFERENCES

1. D. Huang, E. A. Swanson, C. P. Lin, J. S. Schuman, W. G. Stinson, W. Chang, M. R. Hee, T. Flotte, K. Gregory, C. A. Puliafito, and J. G. Fujimoto, "Optical Coherence Tomography," Science **254** (5035), 1178-1181 (1991).
2. B. E. Bouma and G. J. Tearney, *Handbook of Optical Coherence Tomography* (Marcel Dekker, New York, 2002).
3. A. F. Fercher and C. K. Hizenberger, "Optical Coherence Tomography," in *Progress in Optics*, E. Wolf, ed. (Elsevier, North-Holland, 2002), p. 215.

4. Z. Chen, "Functional optical coherence tomgoraphy," in *Frontiers in Biomedical Engineering*, N. H. C. Hwang and S. L.-Y. Woo, eds. (Kluwer Academic/Plenum, New York, 2003), pp. 345-364.
5. Z. Chen, T. E. Milner, D. Dave, and J. S. Nelson, "Optical Doppler tomographic imaging of fluid flow velocity in highly scattering media," Opt. Lett. **22**, 64-66 (1997).
6. Z. Chen, T. E. Milner, S. Srinivas, X. J. Wang, A. Malekafzali, M. J. C. van Gemert, and J. S. Nelson, "Noninvasive Imaging of in vivo blood flow velocity using optical Doppler tomography," Opt. Lett. **22**, 1119-1121 (1997).
7. J. A. Izatt, M. D. Kulkarni, S. Yazdanfar, J. K. Barton, and A. J. Welch, "In vivo bidirectional color Doppler flow imaging of picoliter blood volumes using optical coherence tomography," Opt. Lett. **22**, 1439-1441 (1997).
8. M. D. Kulkarni, T. G. van Leeuwen, S. Yazdanfar, and J. A. Izatt, "Velocity-estimation accuracy and frame-rate limitations in color Doppler optical coherence tomography," Opt. Lett. **23**, 1057-1059 (1998).
9. S. Yazdanfar, M. D. Kulkarni, and J. A. Izatt, "High resolution imaging of in vivo cardiac dynamics using color Doppler," Optics Express **1**, 424 (1997).
10. Z. Chen, Y. Zhao, S. M. Srinivas, J. S. Nelson, N. Prakash, and R. D. Frostig, "Optical Doppler Tomography," IEEE J. of Selected Topics in Quantum Electronics **5**, 1134-1141 (1999).
11. Z. Chen, T. E. Milner, X. J. Wang, S. Srinivas, and J. S. Nelson, "Optical Doppler tomography: imaging in vivo blood flow dynamics following pharmacological intervention and photodynamic therapy," Photochem. Photobiol. **67**, 56-60 (1998).
12. Y. Zhao, Z. Chen, C. Saxer, S. Xiang, J. F. de Boer, and J. S. Nelson, "Phase-resolved optical coherence tomography and optical Doppler tomography for imaging blood flow in human skin with fast scanning speed and high veocity sensitivity," Opt. Letts. **25**, 114 (2000).
13. H. Ren, M. K. Breke, Z. Ding, Y. Zhao, J. S. Nelson, and Z. Chen, "Imaging and quantifying transverse flow velocity with the Doppler bandwidth in a phase-resolved functional optical coherence tomography," Opt. Lett. **27**, 409-411 (2002).
14. U. Morgner, W. Drexler, X. D. Kartner, C. Piltris, E. P. Ippen, and J. G. Fujimoto, "Spectroscopic optical coherence tomography," Opt. Lett. **25**, 111-113 (2000).
15. J. M. Schmitt, S. H. Xiang, and K. M. Yung, "Differntial absorption imaging with optical coherence tomography," J. Opt. Soc. Amer. **A15**, 2288 (1998).
16. M. D. Kulkarni and J. A. Izatt, Conference on Lasers and Electro Optics, Optical Society of America, 59-60, (1996).
17. M. R. Hee, D. Huang, E. A. Swanson, and J. G. Fujimoto, "Polarization-sensitive low-coherence reflectometer for birefringence characterization and ranging," J. Opt. Soc. Amer. B **9**, 903-908 (1992).
18. J. F. de Boer, S. M. Srinivas, A. Malekafzali, Z. Chen, and J. S. Nelson, "Imaging thermally damaged tissue by polarization sensitive optical coherence tomography," Opt. Express **3**, 212-218 (1998).
19. H. Ren, Z. Ding, Y. Zhao, J. Miao, J. S. Nelson, and Z. Chen, "Phase-resolved functional optical coherence tomography: simultaneous imaging of *in situ* tissue structure, blood flow velocity, standard deviation, birefringence, and the Stokes vectors in human skin," Opt. Lett. **27**, 1702-1704 (2002).
20. C. E. Saxer, J. F. de Boer, B. Hyle Park, Y. Zhao, Z. Chen, and J. S. Nelson, "High-speed fiber-based polarization-sensitive optical coherence tomography of in vivo human skin," Opt. Lett. **25**, 1355-1357 (2000).
21. J. Shuliang and L. V. Wang, "Two-dimensional depth-resolved Mueller matrix of biological tissue measured with double-beam polarization-sensitive optical coherence tomography," Opt. Lett. **27**, 101-103 (2002).
22. Y. Jiang, I. Tomov, Y. Wang, and Z. Chen, "Second harmonic optical coherence tomgoraphy," Opt. Lett. **29**, in press (2004).
23. J. S. Nelson, K. M. Kelly, Y. Zhao, and Z. Chen, "Imaging blood flow in human port-wine stain in situ and in real time using optical Doppler tomography," Archives of Dermatology **137**(6), 741-744 (2001).
24. J. F. de Boer, S. M. Srinivas, B. H. Park, T. H. Pham, C. Zhongping, T. E. Milner, and J. S. Nelson, "Polarization effects in optical coherence tomography of various biological tissues," IEEE Journal of Selected Topics in Quantum Electronics **5**, 1200-1204 (1999).
25. M. G. Ducros, J. F. De Boer, H. Huai-En Leah, J. S. Nelson, L. C. Chao, Z. Chen, T. E. Milner, and H. G. Rylander, "Polarization sensitive optical coherence tomography of the rabbit eye," IEEE Journal of Selected Topics in Quantum Electronics 5, 1159-1167(1999).
26. A. F. Fercher, C. K. Kitzenberger, G. Kamp, and S. Y. El-Zaiat, "Measurement of intraocular distances by backscattering spectral interferometry," Opt. Commun. **117**, 43-48 (1995).
27. Y. Zhao, Z. Chen, J. F. de Boer, and J. S. Nelson, "Optical frequency-domain reflectomertry (OFDR) using an integrated fiber tunable filter," Photonic West, Proceedings of SPIE 3598, 56-60, San Jose (1999).

28. S. R. Chinn, E. A. Swanson, and J. G. Fujimoto, "Optical coherence tomography using a frequency-tunable optical source," Opt. Lett. **22**, 340-342 (1997).
29. R. Leitgeb, C. K. Hitzenberger, A. F. Fercher, and M. Kulhavy, "Performance of fourier domain vs. time domain optical coherence tomography," Opt. Express **11**, 889-894 (2003).
30. W. Drexler, "Ultrahigh-resolution optical coherence tomography," J. of Biomedical Optics **9**, 47-74 (2004).
31. Y. Wang, J. S. Nelson, and Z. Chen, "Optimal wavelength for ultrahigh resolution optical coherence tomography," Opt. Express **11**, 1411-1417 (2003).
32. Y. Zhao, Z. Chen, C. Saxer, Q. Shen, S. Xiang, J. F. de Boer, and J. S. Nelson, "Doppler standard deviation imaging for clinical monitoring of in vivo human skin blood flow," Opt. Lett. **25**, 1358-1360 (2000).
33. V. Gusmeroli and M. Martnelli, "Distributed laser Doppler velocimeter," Opt. Lett. **16**, 1358-1360 (1991).
34. Y. Zhao, Z. Chen, Z. Ding, H. Ren, and J. S. Nelson, "Three-dimensional reconstruction of in vivo blood vessels in human skin using phase-resolved optical Doppler tomography," IEEE J. of Selected Topics in Quantum Electronics **7**, 931-935 (2001).
35. Z. Ding, Y. Zhao, H. Ren, S. J. Nelson, and Z. Chen, "Real-time phase resolved optical coherence tomography and optical Doppler tomography," Opt. Express **10**, 236-245 (2002).
36. S. Yazdanfar, A. M. Rollins, and J. A. Izatt, "Imaging and velocimetry of the human retinal circulation with color Doppler optical coherence tomography," Opt. Lett. **25**, 1448-1450 (2000).
37. V. X. Yang, M. L. Gordon, S. Tang, N. E. Marcon, G. Gardiner, B. Qi, S. Bisland, E. Seng-Yue, S. Lo, J. Pekar, B. C. Wilson, and I. A. Vitkin, "High speed, wide velocity dyhamic range Doppler optical coherence tomography (part III): in vivo endoscopic imaging of blood flow in the rat and human gastrointestinal tracts," Opt. Express **11**, 2416-2424 (2003).
38. J. F. De Boer, T. E. Milner, and J. S. Nelson, "Determination of the depth-resolved Stokes parameters of light backscattered from turbid media by use of polarization-sensitive optical coherence tomography," Opt. Lett. **24**, 300-302 (1999).
39. J. Shuliang, Y. Gang, and L. V. Wang, "Depth-resolved two-dimensional Stokes vectors of backscattered light and Mueller matrices of biological tissue measured with optical coherence tomography," Appl. Opt. **39**, 6318-6324 (2000).
40. S. Guo, J. Zhang, L. Wang, J. S. Nelson, and Z. Chen, "Depth-resolved birefringence structure and differential optical axis orientation measurements using fiber-based polarization-sensitive optical coherence tomography," Photonic West, Proceedings of SPIE 3598, in press, San Jose (2004).

Chapter 6

TEMPORAL CLUSTERING ANALYSIS OF CEREBRAL BLOOD FLOW ACTIVATION MAPS MEASURED BY LASER SPECKLE CONTRAST IMAGING

Qingming Luo[1] and Zheng Wang[1]

1. INTRODUCTION

Monitoring the spatio-temporal characteristics of cerebral blood flow (CBF) is crucial for studying the normal and pathophysiologic conditions of brain metabolism. At present there are several techniques for velocity measurement. One of these is laser-Doppler flowmetry (LDF), which provide information about CBF from a limited number of isolated points in the brain (approximately 1 mm^3)[1,2]. Scanning laser-Doppler can be used to obtain spatially resolved relative CBF images by moving a beam across the field of interest, but the temporal and spatial resolution of this technique is limited by the need to mechanically scan the probe or the beam[3,4], such as Laser Doppler Perfusion Imaging(LDPI). Another method is time-varying laser speckle[5-7], which suffered from the same problems as LDF. Single photon emission computed tomography (SPECT) uses the tracer ^{99m}TC-HMPAO to obtain quantitative CBF values (ml/100g/min). However, it suffers from the injection of exogenous substances[8,9]. Positron emission topography (PET) scanning is currently the most versatile and widely used functional imaging modality both in health and disease. The spatial resolution is quite limited, being about 0.5 cm^3 [10,11]. The recently developed thermal diffusion technique is based on the thermal conductivity of cortical tissue, allowing continuous recordings of CBF in a small region of the cortex. The spatial resolution is determined by the placement of the sensor[12,13]. Although autoradiographic methods provide three-dimensional spatial information, they contain no information about the temporal evolution of CBF changes[14]. Method based on magnetic resonance imaging, such as functional Magnetic Resonance Imaging (fMRI) provides spatial maps of CBF but are limited in their temporal and spatial resolution[15,16]. Therefore, a noninvasive simple method removing the need for scanning and providing full-field

---

[1] Correspondence, E-mail: qluo@mail.hust.edu.cn, The Key Laboratory of Biomedical Photonics of Ministry of Education, Huazhong University of Science and Technology, Wuhan 430074, China

dynamic CBF images would be helpful in experimental investigations of functional cerebral activation and cerebral pathophysiology.

One such technique is laser speckle imaging technique (LSI) using the first-order spatial statistics of time-integrated speckle, which is firstly proposed by A. F. Fercher and J. D. Briers[17,18]. They demonstrated that the motion information of the scattering particles could be determined by integrating the intensity fluctuations in a speckle pattern over a finite time. The speckle method has been used to image blood flow in the retina[19] and skin[20]. Lately, the group in Harvard medical school applied this method to image blood flows during focal ischemia and cortical spreading depression (CSD)[21,22].

We developed a modified laser speckle imaging method [23-25] that is based on the temporal statistics of a time-integrated speckle. A model experiment was performed for the validation of this technique. The spatial and temporal resolutions of this method were studied in theory and compared with current laser speckle contrast analysis (LASCA). The comparison indicates that the spatial resolution of the modified LSI is five times higher than that of current LASCA. Cerebral blood flow under different temperatures was investigated by our modified LSI. Compared with the results obtained by LASCA, the blood flow map obtained by the modified LSI possessed higher spatial resolution and provided additional information about changes in blood perfusion in small blood vessels. These results suggest that this is a suitable method for imaging the full field of blood flow without scanning and provides much higher spatial resolution than that of current LASCA and other laser Doppler perfusion imaging methods.

We investigated the spatiotemporal characteristics of changes in cerebral blood volume associated with neuronal activity in the hindlimb somatosensory cortex of a-chloralose-urethane anesthetized rats with optical imaging at 570 nm through a thinned skull[26]. Activation of the cortex was carried out by electrical stimulation of the contralateral sciatic nerve with 5-Hz, 0.3-V pulses (0.5 ms) for 2 s. The stimulation evoked a monophasic decrease in optical reflectance at the cortical parenchyma and arterial sites soon after the onset of stimulation, whereas no similar response was observed at vein compartments. Other group[21] also demonstrated that LSI can monitor the cerebral blood flow (CBF) behaviors with regard to neural mechanisms of brain events under the normal and pathophysiologic conditions, which offered far better spatial and temporal resolution than most alternative imaging techniques thus far. However, it is awkward to show the time course of signals from all spatial loci among the massive dataset especially when we had to deal with thousands of images each of which composes of millions of pixels. Fortunately, temporal clustering analysis (TCA) was proved as an efficient method to analyze functional magnetic resonance imaging (fMRI) data in the temporal domain[31].

In this chapter, we try to use TCA to analyze the data from LSI high-resolution optical imaging. TCA is based on a probability distribution of the overall brain voxels that concurrently reach extreme intensity change of imaging signals, which defined a group of maximal pixels at a certain moment (N_{max}) as a temporal cluster. Mathematically, TCA avoids complicated computation since it converts a multiple-dimension data space into a simple relationship between the number of extremal pixels and the time. And unlike other paradigm-dependent methods it does not require the slightest idea of prior assumptions or knowledge regarding the possible activation patterns. As an example, we presented this novel statistical analysis method to resolve the temporal evolution of evoked CBF changes across somatosensory cortex during sciatic nerve stimulation.

2. PRINCIPLES OF LASER SPECKLE IMAGING

Laser speckle is an interference pattern produced by the light reflected or scattered from different parts of the illuminated rough (i.e., nonspecular) surface. When the area illuminated by laser light is imaged onto a camera, there produced a granular or speckle pattern. If the scattered particles are moving, a time-varying speckle pattern is generated at each pixel in the image. The spatial intensity variations of this pattern contained information of the scattered particles. In areas of increased blood flow, the intensity fluctuations of the speckle pattern are more rapid and the speckle pattern integrated over the CCD camera exposure time becomes blurred in these areas.

To quantify the blurring of the speckles, the local speckle contrast[17,18] is defined as the ratio of the standard deviation to the mean intensity in a small region of the image: $k = \sigma_s/<I>$. Here k, σ_s, and $<I>$ stand for speckle contrast, the standard variation and the mean value of light intensity, respectively. The higher the velocity, the smaller the contrast is. For Gaussian statistics of intensity fluctuations the speckle contrast lies between the values of 0 and 1. A speckle contrast of 1 demonstrated there is no blurring of speckle, therefore, no motion, whereas a speckle contrast of 0 indicates the scatterers are moving fast enough to average out all of the speckles.

The speckle contrast is a function of the exposure time, T, of the camera and is related to the autocovariance of the intensity temporal fluctuations in a single speckle[39], $C_t^{(2)}(\tau)$, by

$$\sigma_s^2(T) = \frac{1}{T}\int_0^T C_t^{(2)}(\tau)d\tau \quad (1)$$

$C_t^{(2)}(\tau)$ is defined as follows:

$$C_t^{(2)}(\tau) = <(I(t)-<I>_t)(I(t+\tau)-<I>_t)>_t$$

where $I(t)$ is the intensity at time t, τ is the "lag", $<>_t$ is the time average.

The normalized autocorrelation function of a field can often be approximated by a negative exponential function (for the case of a Lorentzian spectrum, for example, it is exactly negative exponential[17]):

$$g_t^{(1)}(\tau) = \exp(-\frac{|\tau|}{\tau_c}) \qquad (2)$$

where τ_c is the "correlation time." $g_t^{(1)}(\tau)$ is defined as follows [15]:

$$g_t^{(1)}(\tau) \equiv \frac{\langle E^+(t)E^-(t+\tau)\rangle_t}{\langle I \rangle_t},$$

where $E(t) = E^+(t) + E^-(t)$ is the field at time t and $I(t) = E^+(t)E^-(t)$ is the intensity.

The Siegert relationship[17] is valid for the speckle fluctuations (strictly true only for Gaussian statistics):

$$g_t^{(2)}(\tau) = 1 + |g_t^{(1)}(\tau)|^2. \qquad (3)$$

$g_t^{(2)}(\tau)$ is the normalized second-order autocorrelation function, i.e., the autocorrelation of the intensity, and is defined as follows:

$$g_t^{(2)}(\tau) \equiv \frac{\langle I(t)I(t+\tau)\rangle_t}{\langle I \rangle_t^2}.$$

From the definition of the various correlation functions we have, assuming stationarity:

$$g_t^{(2)}(\tau) = 1 + c_t^{(2)}(\tau), \qquad (4)$$

where $c_t^{(2)}(\tau)$ is the normalized autocovariance,

$$c_t^{(2)}(\tau) = C_t^{(2)}(\tau)/<I>_t^2. \qquad (5)$$

Combining equations 3 to 5 we get:

$$C_t^{(2)}(\tau) = <I>_t^2 |g_t^{(1)}(\tau)|^2. \qquad (6)$$

Assuming our negative exponential approximation for the normalized autocorrelation function, we combine equations 2 and 6 to get:

$$C_t^{(2)}(\tau) = <I>_t^2 \exp(-2\tau/\tau_c)d\tau. \qquad (7)$$

Substituting this expression in equation 1 we obtain the following expression for the spatial variance in the time-averaged speckle pattern:

$$\sigma_s^2(T) = (<I>_t^2 \tau_c/2T)[1 - \exp(-2T/\tau_c)]. \qquad (8)$$

Assuming ergodicity, we can replace the time average by the ensemble average to obtain:

$$k = \frac{\sigma}{\langle I \rangle} = \{\frac{\tau_c}{2T}[1 - \exp(-\frac{2T}{\tau_c})]\}^{\frac{1}{2}}. \qquad (9)$$

Equation 9 gives us an expression for the speckle contrast in the time averaged speckle pattern as a function of the exposure time T and the correlation time $\tau_c = 1/(ak_0v)$, where v is the mean velocity of scatterers, k_0 is the light wavenumber, and a is a factor that depends on the Lorentzian width and scattering properties of the tissue[40]. As in laser-Doppler measurements, it is theoretically possible to relate the correlation times, τ_c, to the absolute velocities of the red blood cells, but this is difficult to do in practice, inasmuch as the number of moving particles that light interacted with and their orientations are unknown[40]. However, relative spatial and temporal measurements of velocity can be obtained from the ratios of $2T/\tau_c$ that is proportional to the velocity and defined as measured velocity in present chapter.

2. METHODS

All animal experiments were performed with the guidelines for neuroscience research[38]. Ten adult male Sprague-Dawley rats (350-400 g, Animal Research Center in Hubei, China) were initially anesthetized with 2% halothane. The right femoral vein was cannulated for drug administration and the right femoral artery for measurement of mean arterial blood pressure (PC Lab Instruments, China). A tracheotomy was executed to enable mechanical ventilation with a mixture of air and oxygen (20% O_2, 80% N_2, TKR-200C, China). Periodically blood gases/acid number were analyzed and kept at physiological arterial blood levels of PaO_2, $PaCO_2$ and pH (JBP-607, Dissolved Oxygen Analyzer, China). The animals were mounted in a stereotaxic frame, and rectal temperature was maintained at 37.0±0.5°C with a thermostatic heating blanket. Anesthesia was continued with an intraperitoneal injection of α-chloralose and urethane (50 and 600 mg/kg, respectively). The skull overlying the hindlimb sensory cortex (2.46×3.28 mm) caudal and lateral to the bregma[10] was bored to translucency with a saline-cooled dental drill, as shown in Figure 1A. Supplemental doses (one-fifth initial dose) were administered hourly, and atropine (0.4 ml/kg sc) was needed to reduce mucous secretions during surgery. The contralateral sciatic nerve was dissected free and cut proximal to the bifurcation into the tibial and peroneal nerves. Then the proximal end was placed on a pair of silver electrodes and bathed in a pool of warm mineral oil.

The speckle contrast imaging instrument was shown as Figure 1B. The single sciatic nerve on the left was stimulated 2 s with rectangular pulses of 0.5 ms duration, 350 mV intensity, and 5 Hz frequency (Multi Channel Systems, Germany). In all animals, a single-trial procedure was repeated 15 to 20 times and separated by an interval of at least 4 min. 400 frames of raw images were obtained in one 10 s single-trial while the electrical stimuli started at 2 s while the images in the first two seconds were recorded as baseline. Images were acquired through the Easy-control software (PCO computer optics, Germany) at 40 Hz and synchronized with Multi Channel Systems. Notably, here data acquisition was synchronized with the electrical signal via an appropriate trigger circuit, and therefore the procedures of data analysis described below could improve the reproducibility of our results and enhance the signal-to-noise ratio[27].

3. RESULTS

With LSI technique we monitored blood flow in somatosensory cortex in a total of 16 rats under electrical stimulation of sciatic nerve, and obtained the activated blood flow distribution at different levels of arteries/veins and the change of activated areas. Although there existed slight differences in individual anatomic features in the rat cortex, we could eliminate this influence since the imaged area was much bigger than the scope demarcated by Hall et al.[41]. One example of our results is shown in Figure 2, in which the brighter areas correspond to the area of increased blood flow. In comparison with LDF, an area of 1 mm^2 ROI in Figure 2 (a) was chosen to evaluate its mean velocity (Figure 3): the evoked CBF started to increase (0.7±0.1) s, peaked at (3.1±0.2) s and then returned to the baseline level. It is coherent with the conclusions obtained from LDF technique[32,34]. In order to differentiate the response patterns of artery/vein under the same stimulus, we labeled six distinct levels of vessels in Figure 2 (a) and displayed their changes of blood flow. The results clearly showed that the response patterns of arteries and veins in the somatosensory cortex were totally different: vein 1 (V-1, ~140 μm in diameter) almost remained unaffected, and arteriole 1 (A-1, ~35 μm in diameter) responded slowly; arteriole 2 (A-2, ~35 μm in diameter) peaked at (3.5±0.5) s after the onset of stimulation and then reached the steady-state plateau, and vein 2 (V-2, ~70 μm in diameter) presented a delay and mild response; blood flow in the capillaries (A-3 and V-3, ~10 μm in diameter) surged readily and increased significantly. We also measured the changes in arteries and veins with different diameters and the results are shown in Figure 4. The statistical results exhibited that arterioles (A-II, ~35 μm in diameter) dilated abruptly ($p<0.05$) but arteriole 1 (A-I) did not change and dilated slightly at 5 to 6 s after the end of stimulation ($p<0.05$). No alterations in vein with diameter of >70 μm were observed during sciatic nerve stimulation ($p>0.05$). We found that the blood flow in capillaries in hindlimb with diameter of >70 μm began to respond at (2.5±0.5) s, dilated up to maximum at (3.5±0.5) s and came back to the prestimulus level; and finally the activation propagated to the entire scope of somatosensory cortex. Blood flow in arteriole 1 did not increase until after 5 to 6 s end of stimulation since it was situated farther from the hindlimb cortex. The activation pattern of cerebral blood flow is discrete in spatial distribution and highly localized in the evoked cortex with the temporal evolution. This is consistent with the hypothesis of Roy and Sherrington and the conclusions drawn by other research groups [32,34]

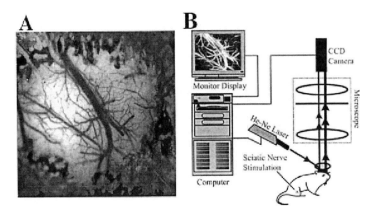

Figure 1. (A) Location of active region. (B) Schematic of system for laser speckle contrast imaging. The light beam from a He-Ne Laser (λ=632.8nm, 3mw) was coupled with a stereo microscope (SZ6045, Olympus, Japan) through an 8mm diameter fiber, which was adjusted to illuminate the area of interest evenly and generated images by a CCD camera (Pixelfly, PCO computer optics, Germany).

Each of LSCI images was a data matrix of 480×640 within the field of view of CCD camera 2.46×3.28 mm, yielding spatial resolution of 27μm. Firstly, a speckle-contrast image (Figure 5 C,D) was computed for each raw speckle image (Figure 5B) by the definition of speckle contrast. Here a template with 5×5 pixels was used. Secondly, a relative blood-flow image was obtained according to the relationship between the speckle contrast and correlation time. Each set of 5 speckle-contrast images and multiple CBF measurements from each imaging session were averaged together before further analysis [27]. Thirdly, we took the image sequences to build a multi-dimension data space (x, y, t), tracked the extremal signals on the pixel-wise LSCI time series, and then counted those extremal pixels (N_{max}) at each image time to generate a histogram. Since the distribution of the maximal signals was subject to random noise in a single trial, the above averaging strategy could accommodate inter-individual variations so that the temporal peaks became more reliable and distinct (Figure6A,B). Next, the time window was determined by data fitting near these peaks by a Gaussian model function. Then a Student's t-test between the pixels inside the time window and the controlled resting pixels was applied to create the spatial activation map. Finally, all the labeled locations (those extremal pixels) were superimposed on a vascular topography of the somatosensory cortex (Figure5A) by means of computer software (Matlab, The Mathworks, Inc., Natick, Massachusetts)(Figure7A,B). The speckle image analysis and TCA algorithm have been described in detail by Briers JD, Dunn AK et al. [21,23,28], Gao JH et al.[31] and curtailed here.

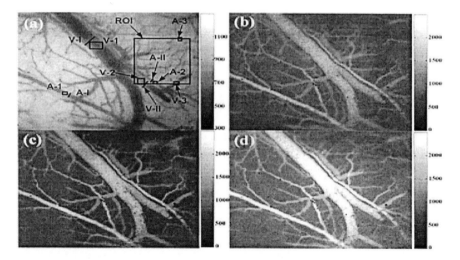

Figure 2. Blood flow change in contralateral somatosensory cortex of rats under unilateral sciatic nerve stimulation. (a) A vascular topography illuminated with green light (540±20 nm); (b)–(d) blood activation map at prestimulus, 1 s and 3 s after the onset of stimulation (the relative blood-flow images are shown and converted from the speckle-contrast images, in which the brighter areas correspond to the area of increased blood flow.), respectively. A-1, 2, 3 and V-1, 2, 3 represent the arbitrarily selected regions-of-interest for monitoring changes in blood flow. A-I,II and V-I,II represent the selected loci on the vessel whose diameters are measured in the experiment.

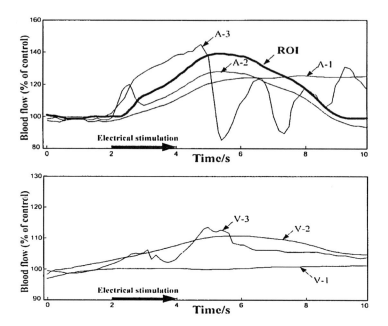

Figure 3. The relative change of blood flow in 6 areas indicated in Figure 2(a) (divided by the values of prestimulus).

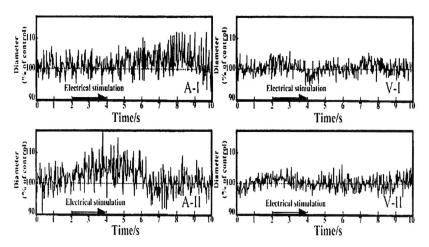

Figure 4. Relative alterations in vessel diameter during sciatic nerve stimulation (divided by the values of presti-mulus)

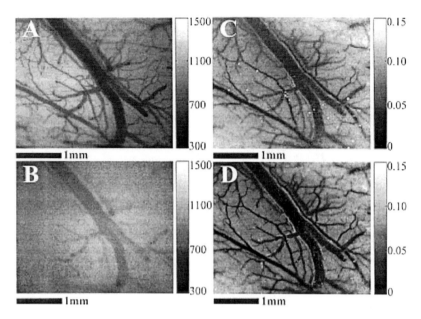

Figure 5. LSCI of a representative animal. (A, B) a vascular topography illuminated with green light (540±20nm) and a raw speckle image with laser. (A) The vascular pattern was referenced in case loss of computation occurred. (C, D) Speckle-contrast images under the prestimulus and poststimulus level demonstrated response pattern of cerebrocortical microflow, in which arteriolar and venous blood flow increased clearly due to sciatic nerve stimulation. The gray intensity bar (C, D) indicates the speckle-contrast values. The darker values correspond to the higher blood flow.

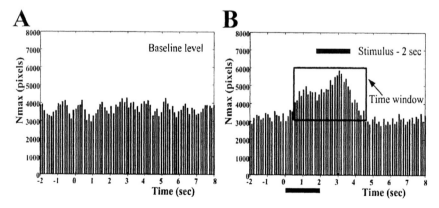

Figure 6. Nmax was a function of time which stood for those pixels that reached a maximal signal at a certain moment and its value on each pixel meant the time when maximal response occurred. (A) Nmax in control experiments without sciatic nerve stimulation, which stood for a baseline level of CBF activities embedded in all physiological and systemic noises (n=6). (B) Temporal maxima of CBF response to sciatic nerve stimulation. A time window (0.50±0.15~4.50±0.10 s) contained a double-peak, 1.10±0.10 s for the first peak and 3.05±0.10 s for the second (n=10). The bold black line represented a stimulus of 2 s duration.

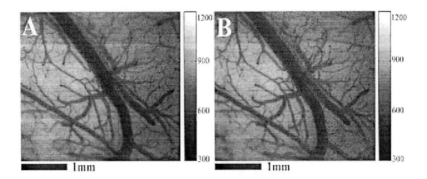

Figure7. Spatial activation map of CBF induced by sciatic nerve stimulation. Two representative images were selected from the relative blood-flow images with labeled the extremal pixels at the double-peak in Figure 3B to display spatial evolution of CBF response across the imaged area. (A, B) Activated locations of CBF at the first and second peak. The dotted areas stood for that changes of CBF reached extrema at the peak moment.

Compared with Figure 5C, these areas of low speckle-contrast values (darker areas) which indicate increased blood flow were clearly evident in Figure 5D. Thus electrical stimulation of the sciatic nerve elicited robust increases in CBF across somatosensory cortex, which was coherent with the former conclusions obtained by LDF[32,34] and fMRI [35]. Since a large temporal and spatial heterogeneity exists in the cerebrocortical vasculature [36], TCA was applied to seek when and where the brain will respond after extern stimuli regardless of individual anatomic features [31].

4. DISCUSSIONS

Temporal and spatial orchestration of neurovascular coupling in the brain neuronal activity is crucial for us to comprehend mechanism of functional cerebral metabolism and pathophysiology. Based on high-resolution optical imaging and TCA techniques, we can extricate ourselves out of traditional paradigm-dependent studies to probe more elaborated spatiotemporal details. The second peak of the double-peak acts in accord with the well-documented results that there possibly exists a rapid release of metabolic factors during the observed latency [32,33]; thus the behavior of regional microcirculation is genuinely adjusted to metabolic activity by chemical regulation of cerebrovascular tone mediated by the products of energy metabolism in the brain. But the mechanism of regulating the first peak in the time window is unclear. It is almost certain that the dynamic regulation of the cerebral microcirculation is mediated by numerous neurogenic and chemical factors acting in concert [29]. Based on the fact that vascular responses to activation are extremely fast [30], neurogenic regulation appears to represent a major role in dynamic control of the first peak. As far as we know, there exist many neural vasomotor pathways in the cerebral vasculature to cause specificities in regulation of blood flow. In view of its spatial distribution localized in capillaries and microarterioles, it is presumably associated with the neurogenic regulation of coupling by classical neurotransmitters such as acetylcholine[37]. Furthermore experiments to investigate neurogenic effects on the cerebral vessels are needed to provide more convincing evidences.

5. ACKNOWLEDGMENTS

This work was supported by grants from National Science Foundation of China for distinguished young scholars (No. 60025514) and National Science Foundation of China (No. 59836240). The authors appreciate Dr. Ngai A.C., Dr. Bandettini P. and Dr. Sokoloff L. for their invaluable help during the preparation of this manuscript. Part of this chapter was published in SPIE proceedings.

6. REFERENCES

1. K. U. Frerichs and G. Z. Feuerstein, "Laser Doppler flowmetry: a review of its application for measuring cerebral and spinal cord blood flow," Molecular Chemical Neuropathology, 12, 55–61(1990).
2. U. Dirnagl, B. Kaplan, M. Jacewicz, W. Pulsinelli, " Continuous measurement of cerebral cortical blood flow by laser-Doppler flowmetry in a rat stroke model, "J. Cereb. Blood Flow Metab. 9, 589–596(1989).
3. B. M. Ances, J. H. Greenberg, J. A. Detre, "Laser Doppler imaging of activation-flow coupling in the rat somatosensory cortex," Neuroimage 10, 716–723(1999).
4. M. Lauritzen and M. Fabricius, "Real time laser-Doppler perfusion imaging of cortical spreading depression in rat neocortex," Neuroreport, 6, 1271–1273(1995).
5. D.A. Zimnyakov, J.D. Briers, and V.V. Tuchin, "Speckle technologies for monitoring and imaging of tissues and tissuelike phantoms" in Handbook of Optical Biomedical Diagnostics PM107, V.V. Tuchin ed. (SPIE Press, Bellingham, 2002), 987–1036.
6. D.A. Zimnyakov and V.V. Tuchin, "Laser tomography" in: Medical Applications of Lasers, D.R. Vij and K. Mahesh eds. (Kluwer Academic Publishers, Boston, 2002), 147–194.
7. E.I. Galanzha, G.E. Brill, Y. Aizu, S.S. Ulyanov, and V.V. Tuchin, "Speckle and Doppler methods of blood and lymph flow monitoring" in Handbook of Optical Biomedical Diagnostics PM107, V.V. Tuchin ed. (SPIE Press, Bellingham, 2002), 881–937.
8 .R. Bullock, P. Statham, , J. Patterson, D. Wyper, D. Hadley, E. Teasdale, " The time course of vasogenic oedema after focal human head injury-evidence from SPECT mapping of blood brain barrier defects". Acta Neurochirurgica (Supplement), 51,286–288(1990)
9. M. Schröder, J. P. Muizelaar, R. Bullock, J. B. Salvant, J. T. Povlishock, "Focal ischemia due to traumatic contusions, documented by SPECT, stable Xenon CT, and ultrastructural studies,"J Neurosurg. 82, 966-971(1995)
10. Alavi, A., Dann, R., Chawluk, J. et al. " Positron emission tomography imaging of regional cerebral glucose metabolism," Seminars in Nuclear Medicine, 16, 2–34 (1996).
11. W. D. Heiss, O. Pawlik, K. Herholz, et al. "Regional kinetic constants and cerebral metabolic rate for glucose in normal human volunteers determined by dynamic positron emission tomography of [18F]-2-fluoro-2-deoxy-D-glucose," Journal of Cerebral Blood Flow and Metabolism, 3, 250–253(1984)
12. L. P. Carter, "Surface monitoring of cerebral cortical blood flow," Cerebrovascular Brain Metabolism Review, 3, 246–261(1991).
13. C. A. Dickman, L. P. Carter, H. Z. Baldwin et al. "Technical report. Continuous regional cerebral blood flow monitoring in acute craniocerebral trauma," Neurosurgery 28, 467–472(1991).
14. Sakurada, C. Kennedy, J. Jehle, J. D. Brown, G. L. Carbin, L, "Sokoloff Measurement of local cerebral blood flow with iodo [14C] antipyrine," Am J Physiol 234: H59–66(1978).
15. D. S. Williams, J. A. Detre, , J. S. Leigh et al. "Magnetic resonance imaging of perfusion using spin inversion of arterial water," Proceedings of the National Academy of Sciences of the USA 89, 212–216 (1992).
16. F. Calamante, D. L. Thomas, G. S. Pell, J. Wiersma, R. Turner, "Measuring cerebral blood flow using magnetic resonance imaging techniques," J Cereb Blood Flow Metab 19, 701–735 (1999).
17. A.F. Fercher and J. D. Briers, "Flow visualization by means of single-exposure speckle photography, " Opt. Commun. 37, 326–329 (1981).

18. J. D. Briers and S. Webster, "Laser speckle contrast analysis (LASCA): A nonscanning, full-field technique for monitoring capillary blood flow," J. Biomed. Opt. 1, 174-179 (1996).
19. K. Yaoeda, M. Shirakashi, S. Funaki, H. Funaki, T. Nakatsue, and H. Abe," Measurement of microcirculation in the optic nerve head by laser speckle flowgraphy and scanning laser Doppler flowmetry, " Am. J. Ophthalmol. 129, 734-739 (2000).
20. B. Ruth, "Measuring the steady-state value and the dynamics of the skin blood flow using the non-contact laser speckle method," Med. Eng. Phys. 16, 105-111 (1994).
21. A. K. Dunn, H. Bolay, M. A. Moskowitz, and D. A. Boas, "Dynamic imaging of cerebral blood flow using laser speckle," J. Cereb. Blood Flow Metab. 21, 195-201 (2001).
22. H. Bolay, U. Reuter, A. K. Dunn, Z. Huang, D. A. Boas, and A. M. Moskowitz, " Intrinsic brain activity triggers trigeminal meningeal afferernts in a migraine model, " Nat. Med. 8, 136-142 (2002).
23. Cheng, H.Y., Luo, Q.M., Zeng, S.Q, Cen, J. and Liang, W.X., Optical dynamic imaging of the regional blood flow in the rat mesentery under the drug's effect, Prog. Nat. Sci., 3(2003) 78-81.
24. Haiying Cheng, Qingming Luo, Shaoqun Zeng, Shangbin Chen, Jian Cen, and Hui Gong. A modified laser speckle imaging method with improved spatial resolution, Journal of Biomedical Optics, 8(3): 559-564, 2003
25. Haiying Cheng, Qingming Luo, Zheng Wang, Hui Gong, Shangbin Chen, Wenxi Liang, and Shaoqun Zeng. Efficient characterization of regional mesenteric blood flow using laser speckle imaging, Applied Optics, 42(28)5759-5764, 2003
26. Pengcheng Li, Qingming Luo, Weihua Luo, Shangbin Chen, Haiying Cheng, and Shaoqun Zeng, Spatiotemporal characteristics of cerebral blood volume changes in rat somatosensory cortex evoked by sciatic nerve stimulation and obtained by optical imaging, Journal of Biomedical Optics, 8(4) 629-635, 2003
27. Ances, B.M., Detre, J.A., Takahashi, K. and Greenberg, J.H., Transcranial laser Doppler mapping of activation flow coupling in the rat somatosensory cortex, Neurosci. Lett., 257 (1998) 25-28.
28. Briers, J.D., Laser Doppler, speckle and related techniques for blood perfusion mapping and imaging, Physiol. Meas., 22(2001) R35-R66.
29. Greger, R. and Windhorst, U., Comprehensive Human Physiology, Vol.1, Springer-Verlag, Berlin, 1996, pp.561-578.
30. Lindauer, U., Villringer, A. and Dirnagl, U., Characterization of CBF response to somatosensory stimulation: model and influence of anesthetics, Am. J. Physiol., 264(1993) H1223-1228.
31. Liu, Y.J., Gao, J.H., Liu, H.L and Fox, P.T., The temporal response of the brain after eating revealed by functional MRI, Nature, 405(2000) 1058-1062.
32. Matsuura, T. and Kanno, I., Quantitative and temporal relationship between local cerebral blood flow and neuronal activation induced by somatosensory stimulation in rats, Neurosci. Res., 40(2001) 281-290.
33. Ngai, A.C., Ko, K.R., Morii, S. and Winn, H.R., Effects of sciatic nerve stimulation on pial arterioles in rats, Am. J. Physiol., 269(1988) H133-139.
34. Ngai, A.C., Meno, J.R. and Winn, H.R., Simultaneous measurements of pial arteriolar diameter and Laser-Doppler flow during somatosensory stimulation, J. Cereb. Blood Flow Metab., 15(1995) 124-127.
35. Silva, A.C., Lee, S.-P., Yang, G., Iadecola, C. and Kim, S.-G., Early temporal characteristics of cerebral blood flow and deoxyhemoglobin changes during somatosensory stimulation, J. Cereb. Blood Flow Metab., 20(2000) 201-206.
36. Steinmeier, R., Bondar, I., Bauhuf, C. and Fahlbusch, R., Laser Doppler flowmetry mapping of cerebrocortical microflow: characteristics and limitations, NeuroImage, 15 (2002) 107-119.
37. Vaucher, E. and Hamel, E., Cholinergic basal forebrain neurons project to cortical microvessels in the rat: electron microscopic study with anterogradely transported Phaseolus vulgaris leucoagglutinin and choline acetyltransferase immunocytochemistry. J.Neurosci., 15(1995) 7427-7441.
38. Zimmermann, M., Ethical principles for the maintenance and use of animals in neuroscience research, Neurosci. Lett., 73(1987) 1
39. Goodman JW , "Some effects of target-induced scintillation on optical radar performance," Proc. IEEE 53, 1688–1700 (1965).
40. Bonner R, Nossal R," Model for laser Doppler measurements of blood flow in tissue," Applied Optics 20, 2097–2107(1981).
41. R.D. Hall, and E.P. Lindholm, "Organization of motor and somatosensory neocortex in the albino rat," Brain Res., 66, 23-28(1974).

Chapter 7

PHOTO- AND SONO-DYNAMIC DIAGNOSIS OF CANCER MEDIATED BY CHEMILUMINESCENCE PROBES

Da Xing[1] and Qun Chen[1]

1. INTRODUCTION

Photodynamic therapy (PDT) is a famous approach in the treatment of cancer. PDT involves the selective uptake and retention of photosensitizers in a tumor, followed by irradiation with light, thereby initiating tumor necrosis through formation of ROS and free radicals in the irradiated tissues.[1, 2] Lots of researches demonstrated that the destroying effect is mainly due to the formation of singlet oxygen ($^1O_2$) resulting from the interaction of an excited photosensitizer with molecular oxygen.[3] Up to now, the evaluation of the effect of PDT is by the measurement of the singlet oxygen production efficiency in vitro. In this paper, we report a rapid imaging method for oxygen free radicals produced in PDT *in vivo* using FCLA (Fluoresceinyl *Cypridina* Luminescent Analog) chemiluminescence to evaluate the efficiency of PDT.

Umemura et al.[4,5] have reported the synergistic effect of ultrasound and hematoporphyrin or gallium-porphyrin derivative ATX-70 on tumor treatment. This phenomenon has been termed "Sonodynamic therapy (SDT)" compared with Photodynamic therapy. SDT appears to be a promising modality for cancer treatment since appears ultrasound can penetrate deep within the tissue and can be focused in a small region of tumor to chemically activate relatively non-toxic molecules thus minimizing undesirable side effects.[6] The clinic application broadening and improvement of SDT efficiency were hindered because the mechanism of sonodynamic action is not clear.

The mechanism of sonodynamic action was said to involve the photoexcitation of the sensitizer by SL light.[5] But Misik and Riesz[6] have shown the evidence against this viewpoint soon later. Sakusabe et al.[7] reported the sonosensitizer increase the antitumor

---

[1] Correspondence, email: xingda@scnu.edu.cn. Tel: +86-20-8521-0089; fax: +86-20-8521-6052. Institute of Laser Life Science, South China Normal University, Guangzhou 510631, China

effects of ultrasound by increasing the production of singlet oxygen ($^1O_2$) and other oxygen species. Yumita et al.[8] suggest that ultrasonically generated active oxygen plays a primary role in ultrasonically induced cell damage in the presence of photofrin II. But Miyoshi et al.[9] have shown the evidence against $^1O_2$ formation by sonolysis of aqueous oxygen-saturated solutions of hematoporphyrin and rose Bengal by use of electron paramagnetic resonance (EPR) spectra of DRD156. Whether $^1O_2$ was involved in the sonosensitization process is still a controversy.

In this chapter, a novel method of photodynamic diagnosis (PDD) of cancer mediated by chemiluminescence probe is presented. The mechanism for photodynamic therapy (PDT) involves singlet oxygen ($^1O_2$) generated by energy transfer from photosensitizers. $^1O_2$ can react with a *Cypridina luciferin* analogue (FCLA), which is a specific chemiluminescence probe for detecting $^1O_2$ and superoxide (O_2^-). The reaction of FCLA and $^1O_2$ can give emission with peak wavelength at about 532 nm. In the present study, FCLA was chosen as an optical reporter of $^1O_2$ produced from photosensitization reaction of Hematoporphyrin Derivative (HpD) in model solution and in nude mice with transplanted mammary cancer. Photosensitized chemiluminescence from the reaction of FCLA with $^1O_2$ was detected by a highly sensitive ICCD system. The chemiluminescence was markedly inhibited by the addition of 10 mmol/L sodium azide (NaN_3) to the model solution and minor effects were observed at the addition of 10 μmol/L superoxide dismutase (SOD), 10 mmol/L mannitol and 100 μg/mL catalase, respectively, thus indicating that $^1O_2$ generation from photosensitization reaction mainly results in the light emission. Also, the chemiluminescence method was engaged to detect the active oxygen species during sonodynamic action *in vitro* and *in vivo*. We used FCLA to real-timely detect oxygen free radical formation in the sonosensitization of HpD and ATX-70. The results clearly show that $^1O_2$ is mainly involved in the sonosensitization. *In vivo* experiments, a tumor-imaging method by sonodynamic chemiluminescence probe was established. This method could have potential applications in clinics for tumor diagnosis.

2. MATERIALS AND METHODS

2.1 Reagents

Photosensitizers of HpD (100mg/20mL), produced in Beijing Institute of Pharmaceutical Industry, were diluted in 0.01 mol/L phosphate buffer saline (PBS, pH 7.4) solution before experiments. FCLA and MCLA[10,11] purchased from Tokyo Kasei Kogyo Co. Ltd., was dissolved in double-distilled water and stored at -20°C until needed. Cu-Zn superoxide dismutase (SOD, from bovine erythrocytes) and human serum albumin (96%) were obtained from the Sigma Chemical Co. Deuterium oxide (D_2O, 99.9 atom % D, from Aldrich Chemical Company, Inc.) was stored under nitrogen until needed. Sodium azide (NaN_3), mannitol, catalase and other chemicals were all made in China.

In vitro experiments, MCLA and luminol (Sigma) were dissolved in double-distilled water with the concentration 50 μmol/L in stock. ATX-70 was purchased from Toyohakka Kogyo, LTD. Superoxide dismutase (SOD) was from Sigma.

In sonication, the sample volume was maintained at 2 ml. The final concentration of MCLA, luminol, ATX-70, HpD, SOD and NaN_3 were 1 μmol/L, 1 μmol/L, 2 μmol/L,

1 μg/ml, 1 μmol/L and 2 mmol/L, respectively. The sample was gas-saturated with air or nitrogen bubbling before sonication.

In vivo experiments, ATX-70 was injected i. p. to the tumor-bearing mouse 24 hours before measurement. FCLA were dissolved in saline (5μmol/L) and injected subcutaneously to the mouse or muscle tissues 5 min before measurement.

2.2 Experimental setup

The setup was similar in nature to that described in our previous study.[12,13] The system used to record the two-dimensional image of the photosensitized chemiluminescence, was based upon a ICCD detector (model: ICCD-576-s/1, from Princeton Instruments Inc. USA), which was combined with a photographic lens of big number aperture (Nikon 50 mm, f 1.4) and a detector controller (ST-130 controller, from Princeton Instruments, Inc. USA). The detection system was a specially designed gated detection. A 50 W high-pressure mercury lamp (from Bio-Rad Co.) in conjunction with a 550 nm long-pass filter (OG 550, from Coherent Co. USA) was used as the irradiation source for photosensitization reactions.

An ultrasonic transducer was driven by a sinusoidal signal from a function generator (Tektronics, AFG320) and a power amplifier (ENI Co. Ltd, 2100L) to produce a column ultrasound field with a frequency at 40 KHz, which passes through distilled water mixed with the black ink as absorption medium to reach the sample for sonodynamic action. The sonodynamic CL light was detected by a photomultiplier tube (PMT) or a cooled intensified charge coupled device (CCD, Princeton Instruments, ICCD-576-S/1) detector. The band-pass optical filter used is centered at 530 nm with a bandwidth of 30 nm, that was used to selectively detect FCLA chemiluminescence to improve the image contrast. A ST-130 controller controls data acquisition of the CCD and transfers the data into the computer in which the digitized image is processed by WINVIEW software.

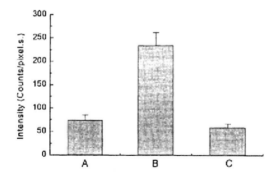

Figure 1. Comparison of photosensitized chemiluminescence intensity of samples. Data are presented as mean ± SD of at least three separate experiments. (A) Light emission from PBS (10 mmol/L, pH 7.4) with HSA (5 μmol/L), HpD (20 μmol/L) and visible light (> 550 nm, 20 s), in the absence of FCLA. (B) Light emission from PBS (10 mmol/L, pH 7.4) with HSA (5 μmol/L), HpD (20 μmol/L), FCLA (2 μmol/L) and visible light (> 550 nm, 20 s). (C) Light emission from PBS (10 mmol/L, pH 7.4) with HSA (5 μmol/L), FCLA (2 μmol/L) and visible light (> 550 nm, 20 s), in the absence of HpD.

All chemiluminescence measurements were carried out at 25°C. Specimen on a tunable stage was placed in a light-tight box. The photocathode was cooled to -40°C to reduce the thermal noise. The spectral response of the detector was 400 nm - 950 nm. The average counts per unit area were calculated in the specimen regions after subtraction of the background counts.

2.3 Preparation of tumor model

Female nude mice with a body weight of about 25 g were used in this study. A saline suspension of 1×10^7 disaggregated Bcap-37 human breast cancer cells were inoculated subcutaneously into the right shoulder of the mice, which caused a tumor to grow at a superficial place. The photosensitized chemiluminescence measurements were performed 2 weeks after inoculation.

3. RESULTS AND DISCUSSIONS

3.1 Measurement of photosensitized chemiluminescence mediated by FCLA in model solution

In experiments, we observed that the model solution contained HpD, HSA and FCLA in PBS showed considerable light emission immediately after the irradiation was stopped and the light emission intensity attenuate rapidly with time, which could last for about 180 s. The photosensitization chemiluminescence intensity are also recorded when either HpD or FCLA was omitted from model solution prior to the performance of irradiation. The comparison of chemiluminescence intensity given in Figures 1 and 2 shows the effect of HpD concentration on the photosensitized chemiluminescence intensity.

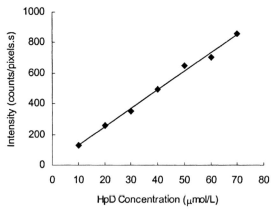

Figure 2. Effect of HpD concentration on the photosensitized chemiluminescence intensity in model solution with PBS (10 mmol/L, pH 7.4), HSA (5 μmol/L), FCLA (2 μmol/L) and visible light (> 550 nm, 20 s).

3.2 Effects of quenchers and D_2O on photosensitized chemiluminescence mediated by FCLA in model solution

FCLA can selectively react with both $^1O_2$ and O_2^- and give chemiluminescence with wavelength at 532 nm.[12] $^1O_2$ and O_2^- can be eliminated by NaN_3 and SOD at appropriate amounts, respectively. In order to confirm the effect of $^1O_2$ on the photosensitized chemiluminescence mediated by FCLA, several quenchers and D_2O were tested in our experiments (results are given in Figure 3). We observed that the addition of the O_2^- scavenger SOD, the H_2O_2 scavenger catalase and ·OH scavenger mannitol to the above model solution prior to the performance of irradiation, did not cause a very significant decrease (decreased by 23.5%, 12.8% and 8.7% respectively compared with control) in the integral chemiluminescence intensity of the first 30 s after irradiation was stopped. However, the integral chemiluminescence intensity was inhibited markedly (about 69.5%) by NaN_3, which is an effective quencher of $^1O_2$. Considering the excellent selectivity of FCLA to $^1O_2$ and O_2^-, the results suggested that the photosensitized reaction of HpD mostly elicit the formation of $^1O_2$ but not O_2^-. (Because of the selectivity of FCLA, in our experiments the generation of ·OH and H_2O_2 in photosensitization action of HpD can not be known well.) The photosensitization reactions of HpD by the substitution of D_2O for H_2O is often used as evidence that $^1O_2$ is involved, since the lifetime of $^1O_2$ is approximately 10 times longer in D_2O than in H_2O, thus allowing it more time to exert effects on targets.[10] The result of photosensitized chemiluminescence in D_2O is also shown in Figure 3.

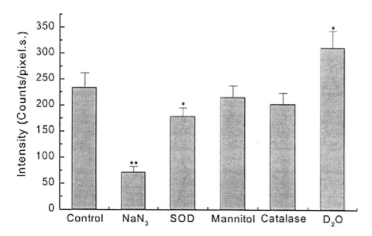

Figure 3. Effects of quenchers and D_2O on photosensitized chemiluminescence mediated by FCLA in model solution. The quenchers and D_2O (99.9 atom % D) were added before irradiation with visible light (> 550 nm, 20 s). Quenchers used are the $^1O_2$ quencher NaN_3 (10 mmol/L), the O_2^- scavenger SOD (10 μmol/L), the ·OH scavenger mannitol (20 mmol/L) and the H_2O_2 scavenger catalase (100 μg/mL). Data are presented as mean ± SD of at least three separate experiments. * $P < 0.05$ in comparison with control; ** $P < 0.01$ in comparison with control.

3.3 Imaging of photosensitized chemiluminescence mediated by FCLA in tumor-bearing nude mouse

When the tumor reached approximately 1000 mm³ in volume after 2 weeks of inoculation, the nude mouse was injected with 200 μg HpD in 0.9% sodium chloride systemically through a tail vein. After a period of 24 hours to allow localization of HpD in the tumor, the mouse was given a superficial injection of 5 μM FCLA of 0.5 mL into the tumor and at the same time as a contrast, the equivalent amount of FCLA was injected subcutaneously in the intact nape of the mouse. Then after a period of 60 minutes for FCLA to diffuse, the mouse was anaesthetized with sodium pentobarbital (50 mg/kg, i.p.). Afterwards the mouse was placed on the black stage in the light-tight box.

The mouse in the light-tight box was irradiated uniformly for 60 seconds with power density as 0.4 W/cm² to allow a photosentization reaction to occur *in vivo*. Photosensitized chemiluminescence mediated by FCLA was recorded immediately after the irradiation was stopped.

Figure 4 shows the images obtained from a photosensitized tumor-bearing nude mouse. The image of (A) in Figure 4 was obtained on illuminating the mouse with dim white light. This was done for the purpose of positioning the mouse and the tumor. The tumor was located in the right shoulder of the mouse. The photosensitized chemiluminescence image recorded for the first 60 seconds after irradiation is shown as (B) in Figure 4. The luminescent area in (B) of Figure 4 corresponds to the tumor region. No clear chemiluminescence was observed around the intact nape and other region of the mouse.

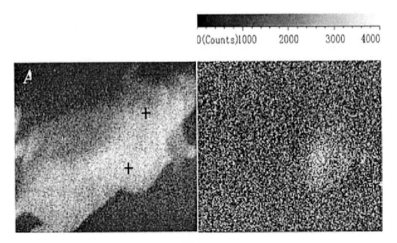

Figure 4. Image of photosensitized chemiluminescence mediated by FCLA in tumor-bearing nude mouse. (A) Image of nude mouse obtained on illumination with dim white light. The tumor can be observed in the right shoulder. Two black crosses indicate the position where FCLA was administered. (B) Photosensitized chemiluminescence image from the tumor-bearing nude mouse. The bright area corresponds to the tumor region. no clear light emission was recorded over the nape. Exposure time is 60 s immediately after irradiation with visible light (> 550 nm, 60 s).

3.4 Measurement of sonosensitized chemiluminescence mediated by MCLA in model solution

Ultrasonic irradiation of liquids generates acoustic cavitation: the formation, growth, and collapse of the bubbles in a liquid creates intense local heating which can drive high energy chemical reactions.[14] Specifically, the sonolysis of water produces ·OH and ·H.[15] These free radicals further form active oxygen spices (in the presence of O_2), which can be detect with CL probe.

A real-time luminescent intensity curve of air-saturated solution is shown in Figure 5. The background counts of the experiments setup are about 40 cps (intensity corresponding to the time 1-4s). The MCLA solution has a luminescent intensity of 650 cps (5-10s) without sonication. When ultrasound is applied, the intensity of MCLA solution increases up to about 12000 cps (11-20s). The emission further great increases to 37000 cps by adding ATX-70 (21-30s), one of the most active sonodynamic agents found. This bright luminescence was inhibited partially by SOD ([down] to 21000 cps during 31-40s) and partially by NaN_3 (further down to 2600 cps during 41-50s).

With another sonosensitizer HpD, we got similar enhancement effect, but the enhancement is from 12000 cps to 28000 cps, relatively smaller than that of ATX-70 (as shown in the left panel of Figure 6). Since the active oxygen spices formation depends on oxygen, we investigated the effect of N2-saturated condition on the CL intensity of the solutions as shown in Figure 6. Under the N2-saturating condition, the CL intensity of both the MCLA solution and the mixture of MCLA and the two sonosensitizer were 3300 cps, significantly lower than that under the air-saturating condition. Since MCLA selectively react with $^1O_2$ or O_2^- to emit photons, from the enhancement of the CL of MCLA solution by sonosensitizer ATX-70 and HpD under ultrasound, one could expect that $^1O_2$ and/or O_2^- was involved during sonication.

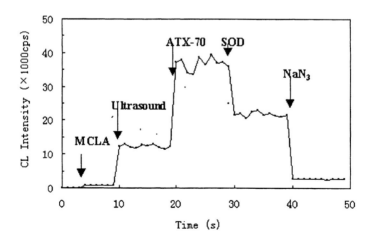

Figure 5. A real-time chemiluminescent intensity curve of air-saturated solution.

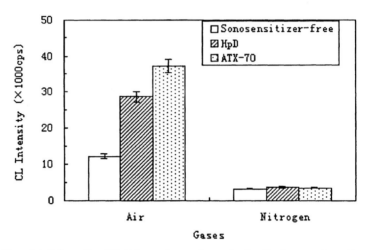

Figure 6. Effect of air (left panel) and N_2 (right panel) saturating on the chemiluminescence intensities of 1 μM MCLA solution without sonosensitizers; MCLA solution in the presence of 1 μg/ml HpD and 2 μM ATX-70 under sonication. Data represent average ± SD from at least three experiments.

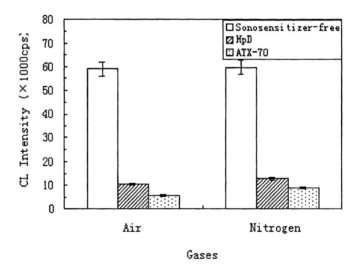

Figure 7. Effect of air (left panel) and N_2 (right panel) saturating on the chemiluminescence intensities of 1 μM luminol solution without sonosensitizers; luminol solution in the presence of 1 μg/ml HpD and 2 μM ATX-70 under sonication. Data represent average ± SD from at least three experiments.

Because the CL was inhibited by NaN_3, a quencher of $^1O_2$, and was not completely inhibited by SOD, $^1O_2$ was expected to be involved in the sonodynamic action. The process of $^1O_2$ formation in sonosensitization need be further investigated.

It is generally accepted that oxidation of luminol by a wide variety of oxidants leads to its characteristic luminescence.[16, 17] Luminol solution can emit strong light during sonication. This bright emission is largely inhibited by adding both HpD and ATX-70 as shown in Figure 7 (left panel). The absorption coefficient of the two sonosensitizers solution with the experimental concentration was 0.011 cm^{-1} and 0.009 cm^{-1} at 580 nm corresponding to the peak wavelength of luminal emission. So light absorption is not enough to explain this inhibition phenomenon. The possible reason may be that sonosensitizer competitively react with the free radicals, which cause luminal emission, produced by cavitation. Figure 7. also shows the CL intensity of luminal solution, both in the presence and in the absence of the two sonosensitizer, was not affected by N_2-saturation. The gas composition does not affect the total free radicals formation, but largely affect the amount of $^1O_2$ and/or O_2^-. So the CL of luminal was caused by the oxidation of ·OH and ·H formed by pyrolysis of water during sonication. ·OH and ·H could react with HpD or ATX-70 to inhibit the CL of luminol solution.

The effect of temperature on air-saturated MCLA solution in the presence of HpD and ATX-70 under sonication was shown in Figure 8. The CL intensity increases linearly with increasing temperature. Because the sonoluminescence intensity increases with decreasing temperature in aqueous solution, the CL, represents the amounts of the $^1O_2$ and O_2^- in the experiments, is not related with the optical excitation of HpD and ATX-70 by the sonoluminescence. With increasing temperature, the sonic impedance of aqueous solution increase, the acoustic energy absorbance increase. This may cause the increasing CL intensity.

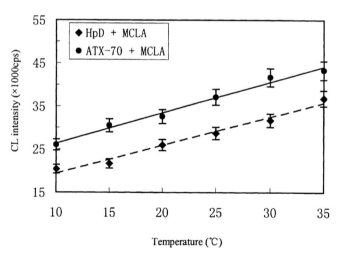

Figure 8. Effect of temperature on air-saturated MCLA solution in the presence of HpD and ATX-70 under sonication. Data represent average ± SD from at least three experiments.

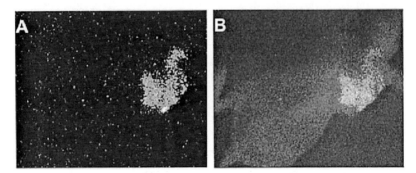

Figure 9. (A) is sonodynamic chemiluminescence image of a tumor bearing nude mouse with ATX-70 and FCLA as sonosensitizer and chemiluminescence probe. (B) is a two-layer image. The bottom layer is a photograph of the mouse obtained on illumination with dim white light. The top layer is the image (A). The bright area corresponds to the tumor region. ATX-70 was injected i. p. 24 hours before. FCLA was injected subcutaneously 5 minutes before measurement.

3.5 Imaging of sonosensitized chemiluminescence mediated by FCLA in tumor-bearing nude mouse

Figure 9 (A) shows the two-dimensional image of sonodynamic CL of ATX-70 mediated by FCLA in a tumor-bearing nude mouse. Figure 9 (B) is a two-layer image for location comparison. The bottom layer of the image is the nude mouse obtained on illumination with dim white light. The tumor can be observed in the right shoulder. The up layer is the image (A). The bright area exactly corresponds to the tumor region, including the shape and the location. The contrast of image is largely improved. In the cases where only ATX-70 or FCLA was injected into the mouse body, there was no significant difference between the emission intensities from the tumor and non-tumor region

FCLA diffused to the tumor tissue within several minutes after it was injected subcutaneously. So the emission from the non-tumor but near the tumor area might be the FCLA-enhanced sonoluminescence of tissues. ATX-70 could accumulate in tumor tissues after it was injected into the body of the mouse 24 hour before measurement, and ultrasound could sensitize ATX-70 to produce singlet oxygen ($^1O_2$) as in vitro experiments, which could react with FCLA to emit chemiluminescence with peak wavelength at 532 nm. So the emission from tumor tissue results from two sources, one is the FCLA-enhanced tissue sonoluminescence, the other is the sonodynamic chemiluminescence.

3.6 The Time-dependence of FCLA retention in cells

Figure 10 showed that FCLA penetrated into cells and localized in prinuclear area. But as one of chemic medicament, there is direct correlation between the retention time of FCLA within cells and causing damage to cells. In the experiment, cells were incubated with 10μM FCLA in complete culture medium at room temperature in the dark for 35 min, the cells were washed with axenic PBS. The average fluorescence intensity of

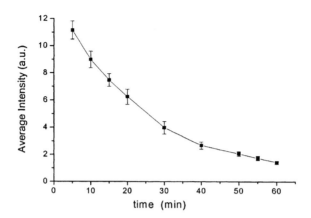

Figure 10. Changes in the intracellular remainder of FCLA fluorescence intensity with time delay. The cells incubated with FCLA in the dark for 35 min.

FCLA in cells for different time was calculated after fluorescence images of FCLA were using confocal microscopy. Figure 10 showed the average fluorescence intensity of retention FCLA within cells decreased with time delay. Area-of-interest analysis revealed 65% loss of FCLA fluorescence by 30 min after FCLA was clear away from extracellular surroundings. ROS, which generated during cellular metabolism, reacted with FCLA and induced decrease of intracellular FCLA. Otherwise, part of FCLA's loss was caused by leakage of FCLA from intracellular space or consumed by its self-oxidation. The average fluorescence intensity of retention FCLA within cells decreased slowly in 40-60 min after FCLA was clear away from extracellular surroundings. Because the retention FCLA within cells decreased, it had little chance of reaction with ROS and amount of FCLA leakage from cells became less. From the result of Figure 10, we found there was litter retention of FCLA within cells when FCLA was clear away from extracellular surroundings after 40-50 min. Because leakage of FCLA from intracellular space or consumed by ROS rapidly, toxicity of FCLA was less and couldn't damage cells because the time of FCLA retention in cell was short less than 60min.

4. CONCLUSIONS

A novel method of photodynamic diagnosis (PDD) of cancer mediated by chemiluminescence probe is presented. The mechanism for photodynamic therapy (PDT) involves singlet oxygen ($^1O_2$) generated by energy transfer from photosensitizers. $^1O_2$ can react with a *Cypridina luciferin* analogue (FCLA) , which is a specific chemiluminescence probe for detecting $^1O_2$ and superoxide (O_2^-). Photosensitized chemiluminescence from the reaction of FCLA with $^1O_2$ at about 530 nm was detected by a highly sensitive ICCD system. The chemiluminescence was markedly inhibited by the addition of 10 mmol/L sodium azide (NaN_3) to the model solution and minor effects were observed at the addition of 10 μmol/L superoxide dismutase (SOD), 10 mmol/L mannitol and 100 μg/mL catalase, respectively, thus indicating that $^1O_2$ generation from photosensitization reaction mainly results in the light emission. *In vivo* photosensitized chemiluminescence imaging was performed with a tumor-bearing nude mouse, and a clear tumor image was obtained. The photosensitized chemiluminescence can be used as a novel evaluation method of cancer photodynamic therapy, and have potential applications in clinics for tumor diagnosis.

On the other hand, the chemiluminescence method was engaged to detect the active oxygen species during sonodynamic action *in vitro* and *in vivo*. We used FCLA or MCLA to real-timely detect oxygen free radical formation in the sonosensitization of air-saturated solution. Adding sonosensitizer, both HpD and ATX-70, could largely enhance the CL intensity. Nitrogen saturation eliminated the CL of FCLA and MCLA, but not affects the luminal CL. The results clearly show that $^1O_2$ is mainly involved in the sonosensitization, and O_2^- was also proved to be involved during the sensitization process. The CL intensity of FCLA or MCLA linearly increases with increasing temperature of sonosensitizer solution. *In vivo* experiments, a tumor-imaging method by sonodynamic chemiluminescence probe was established. With FCLA as CL probe of $^1O_2$ and O_2^-, a tumor sonodynamic CL image was obtained. This method could have potential applications in clinics for tumor diagnosis.

5. ACKNOWLEDGMENTS

This research is supported by the National Major Fundamental Research Project of China (2002CCC00400), the National Natural Science Foundation of China (60378043), and the Research-Team Project of the Natural Science Foundation of Guangdong Province, China (015012). The authors thank Yonghong He, Yunxia Wu and Juan Wang for their contributions in experiments for this. The authors also thank Xiaoyuan Li for his valuble discussions and comments during this research.

6. REFERENCES

1. A. Casas, H. Fukuda, P. Riley, A. M. Batlle, Enhancement of aminolevulinic acid based photodynamic therapy by adriamycin, *Cancer Lett.* **121**, 105-113, 1997.
2. A. Douplik, A. A. Stratonnikov, V. B. Loshchenov, V. S. Lebedeva, V. M. Derkacheva, A. Vitkin, V. D. Rumyanceva, S. G. Kusmin, A. F. Mironov, E. A. Lukyanets, Study of photodynamic reactions in human blood, *J. Biomed. Opt.* **5**, 338-349, 2000.
3. T. Ito, Cellular and subcellular mechanisms of photodynamic action: the singlet oxygen hypothesis as a driving force in recent research, *Photochem Photobiol.* **28**, 493-508, 1978.
4. N. Yumita, R. Nishigaki, K. Umemura, and S. Umemura, Hematoporphyrin as sensitizer of cell-damaging effect of ultrasound, *Jpn. J. Cancer Res.* **80**, 219-222, 1989.
5. S. Umemura, N. Yumita, R. Nishigaki, Enhancement of ultrasonically induced cell damage by a gallium-porphyrin complex ATX-70. *Jpn. J. Cancer Res.* **84**, 582-588, 1993.
6. Y. He, D. Xing, G. Yan, K. Ueda, FCLA chemiluminescence from sonodynamic action in vitro and *in vivo*, *Cancer Lett.* **182**, 141-145, 2002.
7. N. Sakusabe, K. Okada, K. Sato, S. Kamada, Y. Yoshida and T. Suzuki, Enhanced sonodynamic antitumor effect of ultrasound in the presence of nonsteroidal anti-inflammatory drugs, *Jpn. J. Cancer Res.* **90**, 1146-1151, 1999.
8. N. Yumita, S. Umemura and R. Nishigaki, Ultrasonically induced cell damage enhanced by photofrin II: mechanism of sonodynamic activation, *In Vivo*, **14**, 425-429, 2000.

9. N. Miyoshi, T. Igarashi and P. Riesz, Evidence against singlet oxygen formation by sonolysis of aqueous oxygen-saturated solutions of Hematoporphyrin and rose bengal. The mechanism of sonodynamic therapy, *Ultrason. Sonochem.* **7**, 121-124, 2000.
10. K. Sugioka, M. Nakano, S. Kurashige, Y. Akuzawa, T. Goto, A chemiluminescent probe with a Cypridina luciferin analog, 2-methyl-6-phenyl-3,7-dihydroimidazo[1,2-a]pyrazin-3-one, specific and sensitive for $O2-$ production in phagocytizing macrophages, *FEBS Lett.* **197**, 27-30, 1986.
11. M. Nakano, M. Kikuyama, T. Hasegawa, T. Ito, K. Sakurai, K. Hiraishi, E. Hashimura, M. Adachi, The first observation of $O2-$ generation at real time *in vivo* from non-Kupffer sinusoidal cells in perfused rat liver during acute ethanol intoxication, *FEBS Lett.* **372**, 140-143, 1995.
12. Wang J, Xing D, He Y H, et al. Experimental study on photodynamic diagnosis of cancer mediated by chemiluminescence probe, *FEBS Lett.* **523**, 128-132, 2002.
13. He Y H, Xing D, Tan S C, et al. *In vivo* sonoluminescence imaging with the assistance of FCLA, *Phys. Med. Biol.* **47**, 1535-1541, 2002.
14. C. M. Yow, J. Y. Chen, N. K. Mak, N. H. Cheung, A. W. Leung, Cellular uptake, subcellular localization and photodamaging effect of temoporfin (mTHPC) in nasopharyngeal carcinoma cells: comparison with hematoporphyrin derivative, *Cancer Lett.* **157**, 123-131, 2000.
15. K. S. Suslick, Sonochemistry, *Science*, **247**, 1439-1445, 1990.
16. K. Makino, M. M. Mossoba and P. Riesz, Chemical dffects of ultrasound on aqueous solutions: Formation of hydroxyl radicals and hydrogen atoms, *J. Phys. Chem.* **87**, 1369-1377, 1983.
17. G. Bottu, the effect of quenchers on the chemiluminscence of Luminol and lucigenin, *J. Biolumin. Chemilumin.* **3**, 59-65, 1989.

Chapter 8

BIOLUMINESCENCE ASSAY FOR THE HUMAN CHAPERONE MRJ FACILITATED REFOLDING OF LUCIFERASE IN VITRO

Meicai Zhu, Chenggang Liu, Ying Liu, Yinjing Wang, Tao Chen, Xinhua Zhao, and Yaning Liu Liu [1]

1. INTRODUCTION

Compared with the exactly accomplished Human Genome Project, the Functional Genome Plan is even more arduous and complex. The folding of polypeptide, as well as the assembling and disassembling of oligomeric protein complex, are the hinge steps from which the newly translated proteins transformed to their functional conformation. How these are accomplished constitutes a central problem in biology. Because the unfolded proteins can reach their native state spontaneously in vitro, it had been assumed that the folding and the assembling of newly synthesized polypeptides in vivo can occur spontaneously without the catalysis and the input of metabolic energy. This Long-held view has been revised in recent years owing to the discovery that in the cells the correct folding of many proteins depends on the function of a pre-existing protein machinery--the molecular chaperones (Hartl, F. U., 1996). In the presence of ATP, the chaperones worked in coordination and assisted the nascent polypeptide chains to fold up into steric structures. The correct destiny of new proteins was then guaranted (Frydman, J. and Hartl, F. u., 1996). So, the chaperone study is a zealous topic in Post-Genome era.

2. FUNCTION AND STRUCTURE OF MOLECULAR CHAPERONES

The molecular chaperones were defined as proteins that bind to and stabilize an unstable conformation of another protein, and by controlled binding and releasing, facilitate its correct fate in vivo. Molecular chaperones are involved in a wide range of

---

[1] Clinical Lab, Xidiaoyutai Hospital, 30, Fucheng Road, 100036, Beijing, China, email: KZLYN@sina.com.cn

cellular events and play essential roles in a variety of functions. These functions include assisting in the folding of newly translated proteins, guiding translocation proteins across organelle membranes, assembling and disassembling oligomeric protein complexes, and facilitating proteolysis degradation of unstable portions (Hartl, F.U.1996).

As molecular chaperones were initially identified by their appearance under heat stress conditions, they were named as "Heat shock protein (Hsp)". For example, Hsp70, Hsp60 and Hsp40, etc, were the nomentclaters according to their molecular weight in kilo-dalton. In both eukaryotes and prokaryotes, there are their own homologous chaperones. In bacteria, the Dank, GroEL and DnaJ are corresponding to the Hsp70, Hsp60 and Hsp40 of eukaryotes, respectively.

Molecular chaperones do not contain steric information specifying correct folding; instead, they prevent incorrect interactions within and between non-native polypeptides, thus typically increasing the yield but not the rate of folding reactions. This distinguishes them from the so-called folding catalysts, protein disulphide isomerases and peptidyl-prolyl isomerases. These enzymes accelerate intrinsically slow steps in the folding of some proteins, namely the rearrangement of disulphide bonds in secretory proteins and the cis-trans isomerization of peptide bonds preceding proline residues, respectively (Hartl, F. U., 1996).

In Escherichia coli, different classes of molecular chaperones (Dank, DnaJ, GrPE) function in concert: DnaJ and GrPE act as co-factors for the ATPase activity catalyzed by DanK. Mammalian homologues of DanK (Hsp70) and DanJ have also been isolated and in contrast to the prokaryotes, mammals have many homologues of these genes. The Hsp70 family members share a high degree of sequence conservation, but members of DnaJ-like protein family are structurally diverse, containing different combinations of 1-3 conserved domains. All DnaJ-like proteins contain a characteristic J-domain, which is believed to mediate the interaction with Dank (Hsp70) that regulate ATPase activity. The zinc finger-like domain of DnaJ is involved in binding to denatured protein substrates (Szabo, A., et al, 1996). Regions in the C terminus of the DnaJ-like protein are much less conserved and thought to mediate interactions with polypeptide substrate. MRJ (Mouse Related J-protein) is a human DanJ chaperone homologue, which is highly enriched in central nervous system (Figure1). It could not only suppress the polyglutamine-dependent protein aggregation and the cellular toxicity, but is also able to regulate the ATPase activity Hsp70 (Zhuang, J. Z., 2002).

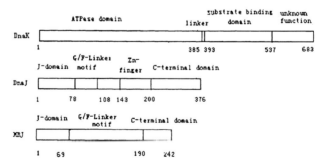

Figure1. Domain organization of DnaK. DnaJ and MRJ. The G/F-linker refers to the G/F amino acid enriched linker.

3. LUCIFERASE AS A MOLECULAR MODEL IN CHAPERONE STUDIES

Using the denatured luciferase as an molecular model for protein refolding and the bioluminescence from liciferin system as an index of luciferase renaturation, the luminescence assay can be taken as one powerful tool for the functional researches of chaperone. Nimmesgern and Hartl analyzed the renaturation of denatured luciferase in the rabbit reticulocyte cystosol. They found that the cystosol contained a highly efficient ATP-dependent protein folding activity, and this protein folding depends on the functional cooperation of different chaperone activities and cofactors in a complex, ATP-dependent process (Nimmesgern, E., Hartl, F. U., 1993). Frydman et al observed the firefly luciferase (62K) synthesis and folding in the rabbit reticulocyte lysate. They found the growing polypeptide interact with a specific set of molecular chaperones (Hsp70, Hsp40, etc). The ordered assembly of these components on nascent chain forms a high molecular mass complex that allows the cotranslational formation of protein domain and the completion of folding once the chain is released from the ribosome (Frydman, J., 1994). The chaperone also plays an important role in the protein translocation across the membrane. The Hsp70 in mitochondria and the ER are not only required for the folding of newly translocated proteins, but also for the translocation process itself. The polypeptide chain traverses the mitochondria membranes through a proteinaceous channel in an extended conformation. Mitochondrial Hsp70 binds to suitable segments of the translating chain as it emerges from the inner membrane, thus preventing it from sliding backwards. Multiple events of Hsp70 binding and ATP-dependent release would then promote translocation in a molecular ratchet-like mechanism (Hartl, F. U., 1996). To investigate roles of DnaJ homologue dj2 and dj1, Terada et al developed a system of chaperone depletion from and readdition to rabbit reticulocyte lysate. They found that heat shock cognate 70 protein (Hsc70) and dj2, but not dj1, are involved in mitochondria import of preornithine trandcarbamylase, and they also tested the effects of these DnaJ homologues in folding of guanidine-denatured firefly luciferase (Terada, K., 1997).

For the first time, we measured the facilitative effects of MRJ on the reactivation of denatured luciferase.

4. EXPERIMENT

4.1 The isolation and identification of MRJ

We isolated and identified MRJ from human skeleton, and from retinal cDNA library as well. Like chaperone Hsp40, MRJ contains a conservative J domain (amino acid 1-69), which is responsible for the distribution of MRJ in the cells.

4.2 The expression and distribution of MRJ in the cells.

4.2.1 The expression of MRJ in COS-7 cell.

Mediated by Lipofectamine, plasmid PMZ10 (full-length MRJ-pRK5) and LY1 (J-domain lacked MRJ-pRK5) were transfeced into COS-7 cells, respectively. The western-blot test showed that the two kinds of target proteins were successfully expressed

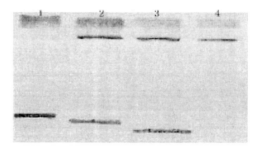

Figure 2. Western-blot of MRJ expressed in the COS-7 cells. 1. Purified MRJ (Positive Control), 2. Extract from COS-7 cells transfected with pMZ10, 3. Extract from COS-7 cells transfected with LY1, 4: Extract from COS-7 cells (Negative Control).

in COS-7 cells (Figure 2, the poitive control was the purified MRJ protein expressed in prokaryocyte. As it contains the histidine chain, so it is longer than the MRJ expressed in COS-7). 24-48 hours after the transfection, the indirect immunofluorescence showed that the exogenous protein, MRJ, was mostly distributed in cytoplasm (Fig .3A). But the MRJ which lacks of J-domain mainly located in the nuclei (Fig 3B)

4.2.2 The expression and distribution of MRJ in CHO cell cycle

The CHO cells were infected with LY2 (full-length MRJ-pcDNA3.1), and then screen the exogenous gene by G418 contained culture medium to obtain the steady expressed CHO cell strain. After double stained by indirect immunofluorescence and Hoechst 33342, the distribution differences of MRJ could be found. The MRJ presented a typical plasma distribution during the interphase (Figure 4). But in the mitotic phase, it appeared an obvious nucleus distribution (Figure 5). The different distribution of MRJ during cell cycle probably suggests that it might be involved in the regulation of cell cycle.

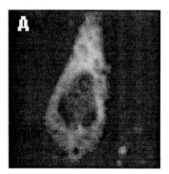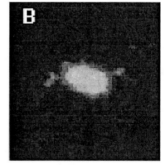

Figure 3. Distribution of MRJ (a) and J-domain defected MRJ (b) in the COS-7 cells

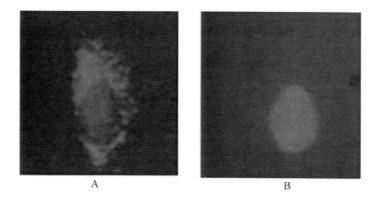

Figure 4. The distribution of MRJ in the interphase CHO cell. A: Green stained for FITC-labeled MRJ; B: Violet-blue stained for Hoechst 33342-binding cell muclei.

4.3. The luminescence assays for the luciferase refolding facilitated by MRJ.

4.3.1. The MRJ expression and protein purification.

Induced by IPTP, MRJ was expressed in E.coli by gene-recombination, and purified by affinity chromatography.

4.3.2. The denature of luciferase

The luciferase (sigma) was chemically denatured in the buffer containing 6mol/L guanidine hydrochloride. The denatured luciferase was used as a molecular model to study the protein refolding and renaturation processes (Terada, K.,1997).

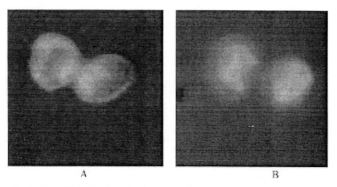

Figure 5. The distribution of MRJ in the mitosis CHO cell. A: Green stained for FITC-labeled MRJ; B. Violet-blue stained for Hoechst 33342-binding cell muclei.

4.3.3. The MRJ facilitated luciferase repatriation verified by bioluminescence

The denatured luciferase was incubated in 25°C with MRJ, Hsp60 and Hsp70 (another two chaperones purchased from Sigma), and different combinations of these three chaperones, respectively. At indicated times, the samples were mixed with bioluminescence system containing Lucifer in, ATP and Mg^{2+} (purchased from Sigma), then their luminosities were measured by KZL-2 luminometer (Stanley, P. E., 1992). In the buffer without chaperones, the luminescence from denatured luciferase and Luciferin, ATP, Mg system was considerable weak and gradually decreased. After the addition of MRJ, the luminosities were somewhat resumed. Higher concentrations of MRJ had more clear effect (Figure 6.).

The concentration of MRJ we used in following experiments was 0.4 μM. When MRJ, Hsp60 and Hsp70 were added into the denatured luciferase solutions respectively, the activities of luciferase all were recovered to different extent and the luminosities from the systems comprehensively increased (Figure 7.).

The combinations of these three chaperones in varying collocations, however, brought about the luminescence further enhanced. Moreover, the bioluminescence of the system was steadily increased along with the prorogated times (Figure 8.). After 60 minutes incubation, the activities of luciferase came to a terrace which corresponding to the 62.8% of the activity of native luciferase (data are not shown).

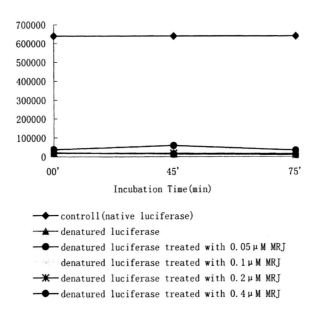

Figure 6. The effects of MRJ in different concentrations on the refolding of denatured luciferase

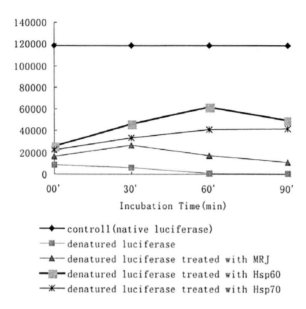

Figure 7. The effects of MRJ, Hsp60 and Hsp70 on the refolding of denatured luciferase

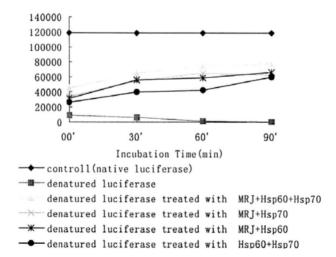

Figure 8. The effects of different combinations of MRJ, Hsp60 and Hsp70 on the refolding of denatured luciferase

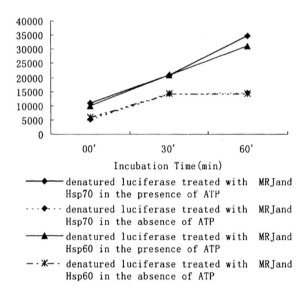

Figure 9. The ATP-dependence of the refolding of denatured luciferase by MRJ, Hsp60 and Hsp70

5. DISCUSSION

Human MRJ is a protein composed of 241 amino acids, and its full-length coding sequence had been cloned. During regulating the ATPase activities,. DanJ-like protein, MRJ, is believed to act as a co-chaperone with Hsp70 protein (Chuang, J. Z., 2002), and the J-domain of DnaJ proteins is thought to regulate the ATPase activity of Hsp70 (Nimmesgern, E., and Hartl, F. U., 1993). Our experiment also shown that the denatured luciferase could not be apparently and steadily reactivated when simply incubated with MRJ (Figure 6.and 7.). While MRJ cooperated with Hsp60 or Hsp70, the reactivations were more effective and affirmative. The combination of these three chaperones gave the best result (Figure 8.).

This fact illustrated that only the assembly of MRJ. Hsp60 and Hsp70 could organize a set of functional synergy mechanism.

The luciferase is one kind of globulin and 37-40% of its steric structure is composed of α-helix. The guanidine denaturant could unfold its α-helix and made it deactivated. The denatured luciferase was chosen as the molecular model for protein refolding and reactivating. Zimmesgern and Hartl enumerated its three advantages: (1) its spontaneous refolding upon dilution from denaturant is inefficient; (2) the luminescence based enzyme assay of luciferase is extremely sensitive, thus allowing to test refolding activities in small fraction volumes; (3) the enzyme assay is very fast so that any reactivation during the measurement is minimal. They also pointed out that the lack of ATP could give rise to the formation of luciferase-chaperones complex which is very difficult to be dissociated, and could lead to the refolding capacity of denatured luciferase decreased (Nimmesgern, E., and Hartl, F. U., 1993). We also found that the refolding and reactivating process was highly ATP-dependent (Figure 9.)

6. ACKNOWLEDGMENT

This work is supported by the National Natural Science Foundation of China, No.69731010.

7. REFERENCES

Chuang, J. Z., Zhou, H., Zhu, M. C., Li, S. H., Li, X. J., Sung, C. H., 2002, Characterization of a brain enriched chaperone, MRJ, that inhibits Hunting aggregation and toxicity independently, J. Biol. Chem., 227(22): 19831-19838.
Frydman, J., Nimmesgern, E., Ohtsuka, K., and Hartl, F. U., 1994, Folding of nascent polypeptide chains in a high molecular mass assembly with molecular chaperones, Nature, 370 (14), 111-117.
Hartl, F. U., 1996, Molecular chaperone in cellular protein folding, Nature, 381: 571-580.
Nimmesgern, E., and Hartl, F. U., 1993, ATP-dependent protein refolding activity in reticulocyte lysate FEBS, 331(1,2): 25-30.
Stanley, P. E., 1992, A survey of more than 90 commercially acailable luminometers and imaging devices forlow-Light measurement of chemiluminescence and bioluminescence, J. Biolumin. Chemilumin., 7: 77-108.
Szabo, A., Korszun, R., Hartl, F. U., and Flanagan, J., 1996, A zinc finger-like domain of molecular chaperone DnaJ is involved in binding to denatured protein substrates, The EMBO Jounal, 15 (2): 408-417.
Terada, K., Kanazawa, M., Bukau, B., Mori, M., 1997, The human DnaJ homologue dj2 facilitates mitochondrial protein import and luciferase refolding, J. cell. Biol., 139 (5): 1089-1095.

Chapter 9

ESSENTIAL DIFFERENCES BETWEEN COHERENT AND NON-COHERENT EFFECTS OF PHOTON EMISSION FROM LIVING ORGANISMS

Fritz-Albert Popp[1]

1. BASIC REMARKS

Biophotons refer to the ultra-weak photon emission from biological systems[1]. The most essential features are displayed in Table 1. The International Institute of Biophysics (IIB) has introduced the term "Biophotonics"[2]. There is general agreement now that Biophotonics concerns the science of using light to understand the inner workings of cells and tissues in living organisms. This provides that spontaneous photon emissions from living tissues[3] (proper biophoton emission), as well as the induced emission of electromagnetic waves from biological systems ("delayed luminescence")[3], are common subjects of scientific investigation and engineering.

One of the most essential questions of biophotonics concerns the coherence of the biophoton field. There are ample indications of a rather high degree of coherence of biophotons [4, 5]. This means that they are providing an essential, if not the most essential information channel in living systems. Consequently, a considerable part of research has to be devoted to the question of coherence and/or incoherence of biophotons.

Table 1. Properties of biophotons

	Universal for all living systems
Emission rate	Continuous photon current rests only after death
Intensity	A few up to some hundred photons/(s.cm^2) $\approx 10^{-17}$ W
Spectral range	At least from 200-800 nm
Spectrum	Continuous, modes are coupled $f(v) \cong$ constant
Photocount statistics	Poissonian, sub- and super-Poissonian
Source(s)	Not yet known with certainty
Polarization	Unknown
Correlations	To **all** biological processes

---

[1] International Institute of Biophysics, 41472 Neuss (Germany), email: iib@lifescientists.de

In most simple terms coherence can be assigned to the capacity of electromagnetic, mechanical or gravitational forces to interfere in space-time in a way that their amplitudes A_1, A_2 ... may affect distinct events by addition or subtraction, depending on whether the forces have the same or opposite directions. This gives rise to interference patterns in space and to significant correlation functions in time.

Since the energies E are proportionate to the square of the amplitudes, the effects follow the law

$$E = (A_1 \pm A_2)^2 = A_1^2 + A_2^2 \pm 2A_1A_2 \qquad (1)$$

Without loss of general validity we take here for simplicity electromagnetic fields. Furthermore, we confine only to two field amplitudes A_1 and A_2, but should mention that the A -values are, in general, rather complicated functions of the position r and the time t. Classically, in case of an electromagnetic field the A-values are subjects of the Maxwell equations. In quantum theory the solutions are probability functions [6] satisfying the Schrödinger-equation (or corresponding equations of quantum field theory). Note that E can take any values between zero and $4A^2$, depending on the values of A_1, A_2, where $0 < |A_1| < |A_2| < |A|$ may be provided in general. Note also that the spatio-temporal distribution E depends strongly on the phases φ of the electromagnetic field strength $A_i = a_i \exp(i\varphi)$, where a_i is the absolute value of the maximum amplitude a (k,ω) after performing a Fourier analysis A(r, t) ⇔ a (k,ω), and φ = (kr-ωt). E (r, t) describes an interference pattern that contains the whole information of the field. For a non-coherent field the phase information breaks down. The incoherence is a result of the random-phase approximation in such a way that the mixed term $2A_1A_2$ of Eq.(1) vanishes after a rather small time interval τ << T. τ is the coherence time and T a relevant observation time of the order of the relaxation dynamics of the system under study. In case of τ<< T, E gets simply a constant value $A_1^2 + A_2^2$. The interference structure disappears then and the field loses its information. Consequently, coherence means in a general sense that the information of the field does not get lost, while incoherence destroys the essential part of spatio-temporal information of the field under study. Coherence provides the highest possible visibility of a structure and the highest possible resolution that can be reached by images of the structure [7]. The visibility is a measure of the contrast; the resolution is the reciprocal of the smallest possible distance that can be resolved.

How important the coherence of electromagnetic fields in biological systems is can be concluded already from a rather famous question of Schrödinger, where an answer is given now in terms of biophotons. Schrödinger asked why the bio-molecules during the cell division do not follow the random Brownian movement but are divided exactly into two parts, spreading out in a non-random way over the daughter cells. We calculated the cavity resonator waves of a typical cell under the constraint of their boundary conditions[8]. We showed that the mitotic figures can be understood as definite superposition of the suitable resonator waves, where the small size of a cell provides that the resonating modes display their eigen frequencies within the optical range (Table 2)[9]. A typical example is shown in Figure1. Consequently, biophotons that have their frequencies just in this optical range provide the most likely answer of the famous Schrödinger question. The extraordinarily long relaxation time of these interference (force-) patterns for guiding the molecular arrangements indicates a rather long coherence time of the superposing electromagnetic field far away from thermal equilibrium.

However, while the spatial structure is subject of the Maxwell equations, the temporal behavior can be only understood in terms of quantum optics[10].

Table 2. Resonator modes of a cavity that has about the same size and properties of a living cell. All these modes are in the optical range. One pattern represents about one biophoton.

TE mode mnp	TM mode mnp	wavelength λ/nm
111		690
	010	574
112		571
	011	546
	012	481
113		462
211		438
	013	410
212		402
114		379
	110	360
213		358
011	111	353
	014	349
012	112	333.5
311		323
115		318

Actually, already a single-frequency solution of the Maxwell equations is generally not a realistic one. A plane monochromatic wave cannot be normalized. Rather, only wave packets with at least some uncertainties in amplitude and phase are realistic physical solutions. The minimum uncertainty wave packets are, for instance, coherent or squeezed states of the Schrödinger equation. Let us understand by x the position, by p the momentum, and by h the Planck's constant. Then we distinguish between two extreme cases, (1) the number state $|n\rangle$ with $\Delta x \Delta p > h$, (2) the coherent state $|\alpha\rangle$ with as well $\Delta x \Delta p = h$ as (after normalization of x and p) $\Delta x = \Delta p$.

For simplicity we refer here at first only to chaotic and coherent states, irrespective of the remarkable fact that in living systems even squeezing ($\Delta x \Delta p = h$, and $\Delta x < \Delta p$ or $\Delta x > \Delta p$) has been shown by experiments[11, 12]. The number state $|n\rangle$ is an eigenstate of the operator $a^+ a$

$$a^+ a |n\rangle = n |n\rangle \tag{2}$$

where n is the number of photons in the field, and a+ and a are the creation and annihilation operators, respectively, satisfying the normalized commutation relation

$$[a, a^+] = 1 \tag{3}$$

The coherent state $|\alpha\rangle$ is defined as an eigenstate of the annihilation operator

$$a |\alpha\rangle = \alpha |\alpha\rangle \tag{4),}$$

where α is the field amplitude.

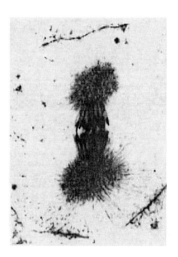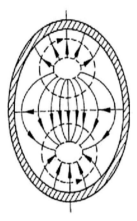

Figure. 1: Left side. Completely developed spindle apparatus of a fish (Corregonus) in mitosis. (From: Darlington, C.D., Lacour, L.F.: The Handling of Chromosomes. Allen and Unwin, London, 1960). Right side. Electic field of TM_{11} cavity modes in a right circular cylindrical cavity. Comparison with Figure 1 left side shows that mitotic figures are striking examples of long-lasting photon storage and coherent fields within biological systems (From: Popp, F.A.: Photon Storage in Biological Systems, In: Electromagnetic Bio-Information, Urban & Schwarzenberg, Muenchen-Wien-Baltimore 1979).

Number states have a definite number of photons, while coherent states display a definite amplitude of the field under study, but follow a completely random (Poissonian) distribution of the photon number.

We cannot go into details here but refer to the textbooks of quantum theory[13]. However, in order to decide whether a field is chaotic or coherent, one has to know the rather refined results of photocount statistics (PCS) which provides the most powerful tool of examining the degree of coherence of an unknown field. This method calculates the probability p (n,Δt) of registering n photons in a preset time interval Δt by recording the number of photons during the measurement time t. One has to distinguish the following cases:

1. $\Delta t \ll \tau_{ch}$, where τ_{ch} represents the coherence time of a *chaotic* field.
2. $\Delta t \gg \tau_{ch}$.

Every chaotic field has a finite (but usually small) coherence time τ_{ch}, whereas the coherence time of a fully coherent field is infinite. The coherence time of ordinary lamps and daylight is of the order of nanoseconds. Laser radiation may arrive at about 100 ms. The coherence time of an optical transition is of the order of the lifetime of the corresponding electronic state. Thus, forbidden optical transitions like radical reactions may display coherence times of the order of some ms.

In a number state the amplitudes of the electric field are fluctuating randomly according to a Gaussian distribution around zero. The energy that is transferred to an ideal photon-detector cumulates up to a photon count rate, where after always definite time intervals Δt a certain photo count number n (Δt) is registered. In view of the mixed term of Eq.(1) that cancels out in a number state sufficiently long after the coherence time, *within* the coherence time the energy E may get at most two times higher than in case of random-phase approximation. As a consequence the probability of registering a second photon once a first one has been registered is for the number state during $\Delta t \ll \tau_{ch}$ at first up to two times higher than for $\Delta t \gg \tau_{ch}$. From exactly the factor 2 at Δt=0 it drops down continuously to 1 for $\Delta t \gg \tau_{ch}$. This effect is called "photon bunching", and it is characteristic for a chaotic field[7]. The exact calculations of p (n,Δt) of a *chaotic field* yield the following results:

1. for $\Delta t \ll \tau_{ch}$,

$$p(n, \Delta t) = \langle n \rangle^n / (1 + \langle n \rangle)^{n+1} \qquad (5a)$$

2. and for $\Delta t \gg \tau_{ch}$,

$$p(n, \Delta t) = \exp(-\langle n \rangle)\left(\langle n \rangle^n / n!\right) \qquad (5b)$$

The first distribution function is known as geometrical distribution, the second the well-known Poissonian distribution.

While a chaotic field expresses by photon bunching some information in the number state representation during its coherence time, this information vanishes completely for time intervals bigger than τ_{ch}.

It may appear rather puzzling that a coherent state displays no information at all in the photocount statistics p(n, Δt). However, this result is just a necessary consequence of quantum theory. Amplitude and photon numbers are not observable with certainty at the same time. Consequently, a sharp field amplitude pattern excludes the measurement of a definitely determined value of the photon number in this state. Thus it is not surprising that the PCS of a fully coherent field is the same as that of a completely chaotic field for $\Delta\tau \gg \tau_{ch}$, i.e. a Poissonian distribution. However, in contrast to a chaotic field the PCS of a fully coherent field is Poissonian for *all* time intervals $\Delta t > 0$ (Figs. 2a and 2b).

The important question comes up whether differences between chaotic and coherent fields display significance at all, that is 1) for $\Delta t \ll \tau_{ch}$, i.e. within the coherence time of the chaotic field, when the chaotic field is classically coherent and the coherent field displays quantum coherence, and when the PCS of chaotic and coherent fields are quite different, and/or 2) for $\Delta t \gg \tau_{ch}$, outside of the coherence time of the chaotic field, when the chaotic field displays no coherence at all and the fully coherent field remains in the state of quantum coherence, while the PCS of chaotic and coherent fields are the same, i.e Poissonian. Does a system react on the different PCS of classical and quantum coherence, or/and does it react on the degree of coherence irrespective of the same PCS?

2. CLASSICAL VERSUS QUANTUM COHERENCE.

In Biophoton-Analysis as well as in Biophotonics, correlation functions play an essential role[6]. They describe the power of the boson fields for transmitting information. In order to distinguish effects of classical and quantum coherence, one has to go back to the roots of quantum theory. Correlation functions of the type

$$\Gamma^{(m,n)}\left(x_1,..,x_{m+n};\tau_2,..,\tau_{m+n}\right) = \lim_{T\to\infty}(1/2T)\int_{-T}^{+T}\prod_{j=1}^{m}V^*\left(x_j,t+\tau_j\right)$$

$$\cdot \prod_{k=m+1}^{m+n} V(x_k, t+\tau_k) dt \qquad (6)$$

where $V(x, t)$ is the "signal", e.g. the electric field amplitude $A(x,t)$ at a point x at time t, are the fundamental quantities.

They are the important measures of characterizing optical patterns (images), for instance the visibility and resolution in space and time. The visibility v corresponds to the contrast of a pattern, introduced by Michelson[14] as

$$v = (I_{max} - I_{min})/(I_{max} + I_{min}) \qquad (7)$$

For

$$|\gamma(x_1, x_2; \tau)| = [\Gamma(1,2,\tau)]/[\Gamma(1,1,0)\Gamma(2,2,0)]^{(1/2)} \qquad (8)$$

we get from

$$I_{max/min} = I_1(x) + I_2(x) + 2[I_1(x) \cdot I_2(x)] \cdot |\gamma(x_1, x_2; \tau)| \qquad (9)$$

COHERENT AND NON COHERENT EFFECTS OF BIOPHOTON EMISSION

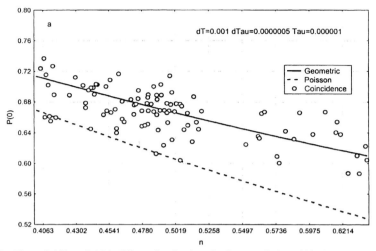

Figure 2a. The probability p(0,Δt) is different for chaotic and coherent radiation within the coherence time of the chaotic light. We measure it by means of coincidence counting[11]. Laser radiation (Stabilized He-Ne-Laser System, Model 05STP903, Melles Griot, Laser Group) is attenuated down to the intensity of biophotons. After passing through a rotating glass plate p(0, Δt) with Δt = 10^{-6} s, the measured p(0) follows rather accurately the theoretical value p(0) of a geometrical distribution. Always thousand values have been averaged for one measurement point. The result shows evidence of the expected chaotic light by random scattering within the coherence time of the laser (measurement points follow the upper curve of the geometrical p(0)).

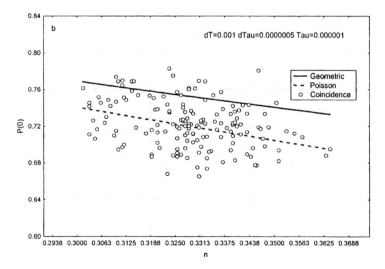

Figure2b. The same light as in Figure2a passes in addition a leaf. The measured p(0) of the penetrating light follows now mainly the theoretical p(0) of a Poissonian distribution. This shows that a biological system may work as a filter that transforms chaotic light within its coherence time into quantum coherent light. An explanation is again given by photon trapping (see text).

just
$$v = |\gamma(x_1, x_2; \tau)| \tag{10}$$
For coherent states we always get
$$|\gamma(x_1, x_2; \tau)| = 1 \tag{11}$$

Chaotic states provide some visibility during their coherence time, but it disappears for $\tau \gg \tau_{ch}$, where τ_{ch} is the coherence time of the chaotic field. The coherence time of thermal light is of the order of nanoseconds, corresponding to the average lifetime of allowed electronic states that are contributing to the radiation. However, a permanent illumination of a subject with pseudo-thermal light is sufficient for an optical image formation within a coherence volume of $(c\tau_{ch})^3$ with c as the velocity of light. It is about 10^4 cm^3. That for "classical" effects quantum theory does not lead to essentially different results may be seen by considering the resolution of optical images, e.g. the smallest distance Δx of an image in a microscope that can be resolved. According to Abbe's theory[14] of a microscope, we have

$$\Delta x \approx \lambda / \sin \theta \tag{12}$$

where $\lambda = \lambda_0 / n$, λ_0 the wavelength of the illuminating light, n the refractive index of the medium, and 2θ the angular aperture of the microscope.

Note that Abbe's theory takes account of interference fringes of the light that may be chaotic or coherent. Consequently, it is based at least on "classical" coherence in terms of a non-vanishing coherence time of the chaotic field.

Considering the same effect in terms of quantum theory, one may calculate Δx by taking account of the uncertainty relation

$$\Delta x \cdot \Delta p \approx h \tag{13}$$

where $p = h/\lambda$ is the momentum of the impinging photons. The uncertainty of p by diffraction into the image plane gets

$$\Delta p = p \sin \theta = (h/\lambda) \sin \theta \tag{14}$$

Insertion into Eq.(13) yields the same result as it has been derived from Abbe's classical diffraction theory. Of considerable importance for Biophoton Analysis, but also for the development of Biophotonics is the fourth order correlation function $\Gamma(2,2)(x_1, x_2; x_1, x_2; 0, -\tau, 0, -\tau) = \langle I_1(t+\tau) I_2(t) \rangle$. It expresses the intensity correlation and it describes the fluctuations of a photon field. A famous example for the importance of this correlation, interferometry of the fourth order, is the Hanbury-Brown-Twiss effect[7]. This correlation function describes the photon bunching of chaotic fields within their coherence time and the lack of bunching of coherent fields and chaotic fields for time intervals $\Delta t \gg \tau_{ch}$. By use of the photon detection equation that describes the process of energy transfer of an electromagnetic field to an ideal photodetector, one obtains from Γ(2,2) the rather basic equation for the variance $\langle(\Delta n)^2\rangle$ of a stationary chaotic field

$$\langle(\Delta n)^2\rangle = \langle n\rangle[1+\langle n\rangle\cdot(\tau_{ch})/(\Delta t)] \qquad (15)$$

within $\Delta t \ll \tau_{ch}$,
while the variance of a fully coherent field follows at any instant

$$\langle(\Delta n)^2\rangle = \langle n\rangle \qquad (16)$$

A chaotic field approaches Eq.(16) in case of $\Delta t \gg \tau_{ch}$.
Since the registration of the photocount rate during time t enables us to calculate the variance $\langle(\Delta n)^2\rangle$ and the mean value $\langle n\rangle$ for every preset Δt with sufficient accuracy, Eq.(15) and Eq.(16) provide a powerful tool of measuring the coherence time of chaotic fields, of examining whether a photon field is chaotic or coherent once the lowest possible limit of τ_{ch} is known.

Apart from the direct application of Eq.(15) and Eq.(16), one may take the underlying PCS (Eq.17) for an even more detailed examination of the field under study.

$$p(n,\Delta t) = \{\Gamma(n+M)/[n!\Gamma(M)]\}(1+M/\langle n\rangle)^{-n}(1+\langle n\rangle/M)^{-M} \qquad (17)$$

where $M = \Delta t/\tau_{ch}$ and Γ represents here the Gamma Function.

Besides direct measurements on Eq.(15) and Eq.(16), we introduced a coincidence of measurements[11, 15-17] with at least two photomultipliers registering the photocount distribution and comparing it with the theoretical values according to Eq.(17) or Eq.(5). It turned out that this method is a rather powerful tool of investigating biophoton fields. I have found this already in wide use among scientists. The results should not be repeated here; however, a few remarks may be useful.

Since radical reactions have life times of the order of 10^{-5} to 10^{-3} s, it turns out, for instance, that biophotons cannot originate from a chaotic field if the hypothesis of radical reactions as the source of biophoton emission were correct. If one takes Δt even smaller than 10^{-5} s, never Eq.(15), but always a result according to Eq.(16) is obtained by experiments. Apart from limiting and exceptional cases this result excludes already random radical reactions as a general source of biophoton emission.

Investigations on dinoflagellates[18], acetabularia and soybeans[19] indicated that biophotons may origin from both, fully coherent and chaotic states, at lowest intensities even from squeezed states[11]. The choice of the (physical) state seems to depend on the biological state. In case of the validity of Eq.(15) the coherence time τ_{ch} has been calculated from Eq.(15) as 0.2 s for soybeans and 1.7 s for dinoflagellates under definite conditions of these systems under study[19].

As a result, this means that the degree of coherence may work as a rather fundamental order parameter for controlling the functions of living systems. This may turn out to provide one of most essential tools of biological regulation. Consequently, one has to ask how the degree (or character) of coherence may organize living matter. I would like to add here a striking example of "photon trapping[20] in living systems" a little more detailed.

Let us take the simple case that the ground state of a system is a number state according to Eq.(2). For simplicity we take the vacuum state $|0\rangle$. Now we switch on a coherent state by adding to $H_0 = fa^+a$ the operator $H_1 = g^*a + ga^+$ and look for a solution of the form $c|0\rangle \pm (1-c^2)^{1/2}|1\rangle$, where the states are number states and c represents the probability amplitude, depending on f and g. The advantage of this approach is that it can be exactly solved and it tells us essential changes in the solution of the problem. A straightforward calculation yields two rather illuminating solutions. First, the system starts to oscillate between the states

$$|t\rangle = c|0\rangle - (1-c^2)^{1/2}|1\rangle \tag{18a}$$

and

$$|e\rangle = c|0\rangle + (1-c^2)^{1/2}|1\rangle \tag{18b}$$

where

$$c^2 = (g^*g)/(g^*g + E^2) \tag{18c}$$

and

$$E_{1,2} = 1/2\left[f \pm (f^2 + 4g^*g)^{1/2}\right] \tag{18d}$$

E has to be replaced by either E_1 or E_2. The difference describes the energy gap that is decisive for the oscillation between the states.

The important insight of this fundamental calculation is that the system can take an energy that is lower than the former ground state energy. This gap provides photon trapping (photon sucking) in a way that oscillations between active photon emission and re-absorption take place (Figure 3). This is the well-known result of a corresponding time-dependent calculation. A more general approach with states $|n\rangle$, where n takes values $n > 1$ shows that the oscillation frequencies v_{osc} follow a law

$$v_{osc} \approx n^\chi \tag{19}$$

where χ is a positive real value around 1/2.

In case of a coherent field, delayed luminescence relaxes according to a hyperbolic (1/t) law. This means that oscillations around the relaxation function are expected[3] that decrease their frequencies according to

$$v_{osc} \approx t^{-\chi} \qquad (20)$$

It should be noted that already in 1981 these kinds of oscillations have been observed[3], and that a first approach pointed in this direction to a solution[21, 22]. Later the existence of these oscillations was several times confirmed, and theories of Dicke[23] and the phase conjugation theory of classical optics[20] indicated the correctness of this first approach. Presently, we have to offer new insight into this rather fundamental phenomenon of living systems that is likely the source of biological clocks, rhythms and communication. Figure 4 demonstrates a striking example. The energy gap of plant tissues has been determined by use of the Arrhenius plot after heating up the system and measuring the amplitude of the oscillation in its dependence of the temperature. We came to a value of about 0.7 eV.

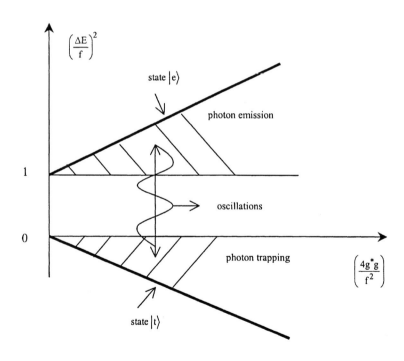

Figure 3. With increasing degree of coherence (g^*g/f^2, see text) an originally chaotic state gets capable to trap ("suck") photons as soon as the Hamiltonian H switches on a coherent part (g^*a+ga^+) such that the total Hamiltonian keeps coherent states coherent. Between ground state and excited states oscillations are induced (see text).

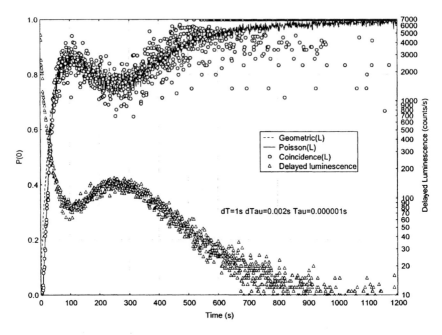

Figure 4. After illumination of a leaf with 780 nm light of a LED, the delayed luminescence displays light oscillations (lower curve with the right scale of the graph in counts/s). The counts show almost accurately the p(0) values of a Poissonian distribution (upper curve p(0) with the left ordinate), but not of a geometrical one, indicating the coherence of the emission. The time interval Δt is here 10^{-6} s. It excludes chaotic radical reactions as a source. A further exclusion is the fact that this phenomenon disappears as soon as one moistures the leaf down to sub-cellular pieces. From a sufficiently low intensity of delayed luminescence (d.l.) on, the statistics get even sub-Poissonian, indicating squeezed light.

3. EXPERIMENTAL SITUATION

There is no doubt any more that spontaneous ultraweak photon emission from living biological systems in the visible range (biophotons) actually exists. It is unlikely within the dimensions of a cell that at least not a considerable degree of coherence becomes decisive in the cellular activity of biophotons. Even for rather small life times of optically allowed electronic transitions the coherence length is of the order of 10^{-9} s times $3 \cdot 10^{10}$ cm/s $\approx$ 30 cm. Over this distance the photons cannot loose their phase information; that means that the cell is always within the coherence volume of the biophoton field even if biophotons were chaotic. It means at least that it is impossible to assign a definite source to biophotons, which are measured outside of the cell. Although it is customary to use correlation functions for determination of a source of photon emission, the fact that one never could determine a definite source of biophoton emission is an indication for this situation. Only squeezed light may help to change the resolution. On the other hand, biophotons are not necessarily products of chemical reactions but may provide with the same or even higher probability the very trigger impulses of chemical reactions.

Consequently, that biophotons are correlated to every function within a cell including the unwinding and upwinding of the DNA by intercalation of Ethidium Bromide[24] is a further strong indication for a coherent field, either within the coherence range of number states or even quantum coherence. Evidence of quantum coherence as an obvious state of the biological photon field is shown by the experimental proof of squeezed light, since it is a necessary pre-step of squeezing. Quantum coherence has been proven by the Poissonian PCS for showing evidence of the ergodic character of this field, and the hyperbolic-like relaxation dynamics of delayed luminescence as a sufficient condition for an ergodic field. Both results belong together in order to show evidence of the quantum coherence of biophotons. Apart from the physical properties, classical and also quantum coherence of biophotons should have considerable impact on biological effects. A lot of papers have already appeared about some not commonly understandable phenomena that could be explained in terms of quantum coherence of biophotons[25]. I would like to mention here some examples: cell growth and differentiation, mediated by intra-and intercellular regulation and communication by coherent photons as well as by photon trapping ("sucking") in cell populations. The only possible explanation of why about 10,000 chemical reactions in a cell could take place always at the right position and at the right time is the coherent (squeezed) excitation of the suitable electronic states of the reactants by biophotons. The reaction volume of a cell is so small and the necessary intensity is as high as to exclude *truthfully* thermal photons as a possible Arhennius-source of chemical reactivity. Photon trapping has been indicated, for instance, by experiments of Galle[26], Scholz[27], Schamhart[28], Vogel[29], Beloussov[30].

This phenomenon, which is theoretically explained above, has to be looked upon as the real source of life, since it describes the only essential and measurable difference between living and non-living matter as far as it is known at present. Photon trapping provides the rather long lifetime of metastable states in biological tissues as well as the permanent competition for photons. At the same time the delocalization of the biophoton field finds it natural explanation. This provides a mechanism of intercellular communication where experimental work has been performed among others by the pioneering work of Alexander Gurwitsch[31] and recently by Albrecht-Bühler[32], Chang et.al.[18], and Shen[33]. Photon trapping provides a basis of oscillatory behaviour of the biophoton field as has been shown already in Figs.3-4. However, this phenomenon can be observed also by measuring the biophoton field of the human body shown by Cohen[34]. A further pioneering step in this development are non-local effects of influences on the skin, where under certain conditions the local increase of biophoton emission gets distributed to completely different parts of the body with a long-time relaxation. This points to light-channels in living systems, as has been demonstrated recently in plant tissues by Mandoli and Briggs[35]. This "light piping in plant tissues" has been assigned to the coherent nature of the photon field. While in healthy people the biophoton field follows right-left symmetry, in sick people asymmetric biophoton emission has been observed[34, 36]. In particular MS-patients display rather high and asymmetric biophoton emission.

The manifold of experimental data from Gurwitsch's area on invites us at present time to intensify the research in the following topics:

(1) Investigation of the spectral distribution. It turned out that the modes are coupled and that this is the way in which biological systems minimize the entropy. Consequently, the degree of coupling between different modes may provide one of the most powerful tools in order to assess health and disease of living systems.

This and similar research is now subject of the Japanese groups under the guidance of Makino.

(2) Investigation of temperature-dependence of biophoton emission and delayed luminescence. The temperature dependence displays rather characteristic temperature-hysteresis curves. Therefore, the degree of cooperative effects becomes observable by measuring the intensity dependent on the temperature change. Also at the same time, spectral shifts and homeostasis become interesting subjects of temperature behavior.

(3) Investigations of the biophoton field of the human body. After Cohen's work as well oscillations, left-right asymmetries and non-local effects of biophotons and delayed luminescence in dependence of health and disease of men may open a completely new door in medical diagnosis and therapy.

(4) Investigations of the degree of coherence in terms of correlation functions and coincidence counting. Measurements of this kind may reveal the importance of biological order parameters like $\Gamma(n,m)$ in terms of Glauber's correlation functions.

A lot more research is needed, e.g. physical and chemical influences on biophotons including delayed luminescence, the dependency on cell growth and differentiation, intracellular mechanisms like transcription and replication (that might be regulated by transformation from quantum coherence to the coherence area of chaotic fields). This may lead to a more basic understanding of biological quality in the most general sense.

4. ACKNOWLEDGEMENTS

The author likes to thank the German Ministery of research for financial support, in particular Wolfgang Klimek, for profound discussions and basic theoretical support, Yu Yan, Zhongchen Yan and Sophie Cohen for tireless help in the experimental work, Roeland van Wijk, Eduard van Wijk, Rajendra Bajpai, Jiin-Ju Chang, Gerard Hyland, Lev Beloussov, Takahira Makino, John Swain, Xun Shen, Kwan-Sup Soh and Vladimir Voeikov for very valuable and stimulating lectures and discussions.

5. REFERENCES

1. J.J.Chang, J.Fisch and F.A.Popp (eds.), *Biophotons* (Kluwer Academic Publishers, Dordrecht, 1998).
2. L.V.Beloussov and F.A.Popp (eds.), *Biophotonics:Non-equilibrium and Coherent Systems in Biology, Biophysics and Biotechnology* (Bioinform Services Co., Moscow, 1995).
3. F.A.Popp, B.Ruth, W.Bahr, J.Böhm, P.Grass, G.Grolig, M.Rattemeyer, H.G.Schmidt, and P.Wulle, Emission of visible and ultraviolet radiation by active biological systems, *Collective Phenomena* (Gordon&Breach) **3**, 187-214 (1981).
4. F.A.Popp and K.H.Li, Hyperbolic relaxation as a sufficient condition of a fully coherent ergodic field, *Intern.J.Theor.Phys.* **32** (9), 1573-1583 (1993).
5. F.A.Popp and Y.Yan, Delayed luminescence of biological systems in terms of coherent states, *Physics Letters A* **293**, 93-97(2002).
6. R.J.Glauber, Coherence and Quantum Detection, in: *Quantum Optics*, edited by R.J.Glauber (Academic Press, New York and London, 1969), pp. 15-56.
7. J.Perina, *Coherence of Light* (van Nostrand Reinhold Company Ltd, London, 1971).
8. F.A.Popp, Photon Storage in Biological Systems, in: *Electromagnetic Bio-Information,* edited by F.A.Popp, G.Becker, H.L.König, and W.Peschka (Urban & Schwarzenberg, Munich-Vienna-Baltimore, 1979), pp.123-149.

9. M.Rattemeyer, Modelle zur Interpretation der ultraschwachen Photonenemission aus biologischen Systemen, Diplomarbeit, Fachbereich Physik der Universität Marburg, 1978.
10. R.J.Glauber, Coherent and incoherent states of the radiation field, *Phys.Rev.* **131** (6), 2766-2788 (1963).
11. F.A.Popp, J.J.Chang, A.Herzog, Z.Yan, and Y.Yan, Evidence of non-classical (squeezed) light in biological systems, *Physics Letters A* **293**, 98-102 (2002).
12. R.Bajpai, Biophoton Emission in a Squeezed State from a Sample of Parmelia.tinctorum, *Physics Letters A* (2004) in press.
13. D.F.Walls and G.J.Milburn, *Quantum Optics* (Springer, Berlin, 1994).
14. M.Born and E.Wolf, *Principles of Optics* (Pergamon, Oxford, 1975).
15. F.A.Popp, J.J.Chang, Q.Gu and M.W.Ho, Nonsubstantial Biocommunication in Terms of Dicke's Theory, in: *Bioelectrodynamics and Biocommunication*, edited by M.W.Ho, F.A.Popp and U.Warnke (World Scientific, Singapore-London, 1994), pp. 293-317.
16. F.A.Popp and Y.Yan, Verfahren zur Ermittlung geringster Qualitätsunterschiede, Patentanmeldung 10147701.5 (9/27/01).
17. F.A.Popp and X.Shen, The photon count statistics study on the photon emission from biological systems using a new coincidence counting system, in: *Biophotons*, edited by J.J.Chang, J.Fisch and F.A.Popp (Kluwer Academic Publishers, Dordrecht, 1998), pp.87-92.
18. J.J.Chang and F.A.Popp, Biological Organization: A possible mechanism based on the coherence of biophotons, in: *Biophotons*, edited by J.J.Chang, J.Fisch and F.A.Popp (Kluwer Academic Publishers, Dordrecht, 1998), pp. 217-227.
19. F.A.Popp, Some Essential Questions of Biophoton Research and Probable Answers, in: *Recent Advances in Biophoton Research and its Applications*, edited by F.A.Popp, K.H.Li and Q.Gu (World Scientific, Singapore-London, 1992), pp. 1-46.
20. F.A.Popp and J.J.Chang, Mechanism of interaction between electromagnetic fields and living organisms, *Science in China* (Series C), 507-518 (2000).
21. W.Nagl and F.A.Popp, A physical (electromagnetic) model of differentiation, 1. Basic Considerations, *Cytobios* **37**, 45-62 (1983).
22. F.A.Popp and W.Nagl, A physical (electromagnetic) model of differentiation. 2. Applications and examples, *Cytobios* **37**, 71-83 (1983).
23. K.H.Li, Coherence in Physics and Biology, in: *Recent Advances in Biophoton Research and its Applications*, edited by F.A.Popp, K.H.Li and Q.Gu (World Scientific, Singapore-London, 1992), pp 113-195.
24. M.Rattemeyer, F.A.Popp and W.Nagl, Evidence of photon emission from DNA in living systems, *Naturwissenschaften* **68** (11), 572-573 (1981).
25. F.A.Popp, Q.Gu and K.H.Li, Biophoton emission: Experimental background and theoretical approaches, *Modern Physics Letters B* **8**, Nos. 21 & 22, 1269-1296 (1994).
26. M.Galle, Population density-dependence of biophoton emission from daphnia, in: *Recent Advances in Biophoton Research and its Applications*, edited by F.A.Popp, K.H.Li and Q.Gu (World Scientific, Singapore-London, 1992), pp. 345-355.
27. W.Scholz, U.Staszkiewicz, F.A.Popp and W.Nagl, Light stimulated ultraweak photon reemission of human amnion cells and Wish cells, *Cell Biophysics* **13**, 55-63 (1988).
28. D.H.J.Schamhart and R.van Wijk, Photon emission and the degree of differentiation, in: *Photon emission from biological systems*, edited by B.Jezowska-Trzebiatowska, B.Kochel, J.Slawinski and W.Strek (World Scientific, Singapore-New Jersey, 1986), pp.137-152.
29. R.Vogel and R.Süßmuth, Weak light emission from bacteria and their interaction with culture media, in: *Biophotons*, edited by J.J.Chang, J.Fisch and F.A.Popp (Kluwer Academic Publishers, Dordrecht, 1998), pp.19-44.
30. L.V.Beloussov, A.B.Burlakov and A.A.Konradov, Biophoton Emission from Eggs and Embryos of a Fish Misgurnus fossilus: Development Dynamics, Frequency Patterns and Non-Additive Interactions, in: *Biophotonics and Coherent Systems*, edited by L.Beloussov, F.A.Popp, V.Voeikov and R.van Wijk, (Moscow University Press, 2000), pp. 305-320.
31. A.G.Gurwitsch and L.D.Gurwitsch, *Die mitogenetische Strahlung, ihre physikalisch-chemischen Grundlagen und ihre Anwendung in Biologie und Medizin* (Gustav Fischer Verlag, Jena, 1959).
32. G.Albrecht-Bühler, Rudimentary form of cellular vision, *Proc.Natl.Acad.Sci.USA* **89**, 8288-8292 (1992).
33. X.Shen, L.Bei, T-H.Hu and B.Aryal, The possible role played by biophotons in the long-range interactions between neutrophil leukocytes, in: *Biophotonics and Coherent Sytems*, edited by L.Beloussov, F.A.Popp, V.Voeikov and R.van Wijk, (Moscow University Press, 2000), pp. 335-346.
34. S.Cohen and F.A.Popp, Whole-body counting of biophotons and its relation to biological rhythms, in: *Biophotons*, edited by J.J.Chang, J.Fisch and F.A.Popp (Kluwer Academic Publishers, Dordrecht, 1998), pp.183-191.

35. D.F.Mandoli and R.W.Briggs, Optical properties of etiolated plant tissues, *Proc.Natl.Acad.Sci.USA* **79**, 2902 (1982).
36. H.-H. Jung, W.-M.Woo, J.-M.Yang, C.Choi, J.Lee, G.Yoon, J.S.Yang, S.Lee and K.S.Soh, Left-right asymmetry of biophoton emission from hemiparesis patients, *Indian J.Exp.Biol.* **41**, 452-456 (2003).

Chapter 10

PARAMETERS CHARACTERIZING SPONTANEOUS BIOPHOTON SIGNAL AS A SQUEEZED STATE IN A SAMPLE OF *PARMELIA.TINCTORUM*

R.P.Bajpai[1]

1.INTRODUCTION

Almost every living system after a few seconds exposure to normal laboratory illumination emits a weak photon signal of unusual features [1,2]. Such a signal is not emitted by a dead system. A widely observed unusual feature is a small decaying portion and a long non-decaying tail in these photon signals. The decaying portion lacks exponential decay character and hence the parameters characterising the decay are difficult to identify. The shape and strength of decaying portion are sensitive to many physiological and environmental factors. Perhaps, both are system and situation specific. The non- decaying tail is observable during the entire lifetime of a system; the living system continues to emit an almost constant flux of photons. The non-decaying portion is observable for hours in a quasi-stable system. The non-decaying portion obviously lacks a decay character whether exponential or non-exponential; which is much easier to ascertain. The strength of this portion is too weak to discern its sensitivity to various physiological and environmental factors. The semi-classical framework is usually employed for describing photon signal. The framework envisages photon emission from the probabilistic decay of subunits in the excited states. A subunit in living systems is likely to be a biomolecule or complex structure made up of many biomolecules and the emitted photon signal contains information about the subunit. The probabilistic decay allows only exponentially decaying photon signals. Lack of exponential decay character implies correlated photon-emitting subunits that need holistic description[3]. A description based on independent bio molecular structures will be erroneous and could lead to paradoxes. Adding the prefix bio to these photon signals and associated photons indicates the need of a holistic framework and biological connection. A photon signal of unusual

---

[1] ISOSB, North Eastern Hill University, Shillong 793022, India; rpbajpai@nehu.ac.in and International Institute of Biophysics, IIB e.V. ehem. Raketenstationen, Kapellener Straße, D-41472 Neuss, Germany. (Tel: 0091- 364 -2550299)

features emitted by a living system is called biophoton signal and its photons are called biophotons. A biophoton signal has many unusual properties but we shall concentrate on shape, situation specific nature and detection probabilities of different number of photons; these properties suggest the signal to be a photon signal in a squeezed state [3]. It is pointed out that the unusual features of the non-decaying portion were first identified and that of the decaying portion were identified after a while. As a result, the two portions are still distinguished, the non-decaying and decaying portions are identified respectively as spontaneous and stimulated or light induced biophoton signals.

Popp and Li [4] proposed a phenomenological model for explaining non-exponential decay of signals. The model envisages dynamical origin of the shape and is successful in reproducing the broad features of light induced biophoton signals. The model postulates that the following effective Hamiltonian H describes the dynamics of a biophoton field:

$$H = \frac{p^2}{2(1+\lambda t)^2} + \frac{1}{2}(1+\lambda t)^2 \omega^2 q^2 \qquad (1.1)$$

where λ is damping coefficient, ω is mode frequency, t is time, and p and q are canonically conjugate quadratures of free photon field. The Hamiltonian corresponds to a frequency stable damped harmonic oscillator with time dependent damping and mass terms. The ensuing classical equation of motion has the following analytic solution:

$$q = \frac{q_0}{(1+\lambda t)} \sin(\omega t + \varphi) \qquad (1.2)$$

where q_0 and φ are integrating constants and are determined by the initial state of the system or initial conditions. The amplitude of the oscillator decreases hyperbolical with time. The decrease of energy of the oscillator averaged over the mode frequency is given by

$$\langle H \rangle = \frac{1}{2} \omega^2 q_0^2 + \frac{1}{4} \frac{\lambda^2 q_0^2}{(1+\lambda t)^2}. \qquad (1.3)$$

It gives the shape of the classical signal that decays non-exponentially in time. The expression of energy given in eq. (1.3) is problematic for visible range photons having $\omega \approx 10^{16}$ rad/s, which makes the constant term of the expression much larger than its time varying term and the decay of the signal in the visible range unobservable. The problem is circumvented by assuming that the constant term determines the threshold voltage of detecting photons and the time varying term determines the shape of the signal. The canonical parameters specifying a biophoton signal are λ and q_0 and not strength and decay constant. The strength and shape can still be defined operationally; strength by the number of photo counts observed in a definite interval at some fixed time and shape by the relative distribution of photo counts at different times. The strength thus defined depends on initial conditions and not the shape, so that strength will be situation specific and not the shape. The success of the model in reproducing the observed data is rather poor but improves dramatically if decay exponent 2 is replaced by a situation specific ad-hoc parameter m. The shape of the signal is then given by $(1+\lambda t)^{-m}$. The value of m lies between 1 and 2 in different biophoton signals. The ad-hoc replacement is difficult to incorporate in the classical framework and is attributed to quantum effects.

The quantum dynamics of the Hamiltonian H has analytic solution[5] as well. The quantum solution is either a coherent state $|\alpha(t)\rangle$ or squeezed state [6] $|\alpha(t), \xi(t)\rangle$ of photons. The coherent or squeezed state solution depends upon the choice of initial state. The Hamiltonian H gives the time evolution of both states and the evolution preserves the coherent or squeezed nature. The photon signal obtained from a coherent state decays only to half of its peak value and is not suitable for describing biophoton signals decaying to 2 to 3 orders of smaller values. The coherent state solution is, therefore, ignored. The squeezed state solution can describe the behaviour of observed biophoton signals. Two complex parameters $\alpha(t)$ and $\xi(t)$ specify a squeezed state solution and their time dependences specify the evolution of state. These parameters specify displacement and squeezing required for obtaining the squeezed state solution from the vacuum state $|0\rangle$ by the application of displacement operator D $(\alpha(t))$ and squeezing operator S $(\xi(t))$ i.e.

$$|\alpha(t), \xi(t)\rangle = D(\alpha(t))S(\xi(t))|0\rangle. \tag{1.4}$$

The displacement and squeezing operators are well known operator functions of annihilation and creation operators a and a^+ of free photon field:

$$D(\alpha) = \exp(\alpha a^+ - \alpha^* a) \text{ and } S(\xi) = \exp\left[\frac{1}{2}\left(\xi^* a^2 - \xi a^{+2}\right)\right]. \tag{1.5}$$

The analytic expressions specifying the dependence of $\alpha(t)$ and $\xi(t)$ on time, initial conditions and damping have been calculated explicitly, which allows the calculation of various measurable quantities and their dependence on time, initial conditions and damping e.g. expectation value of photon number operator a^+a averaged over the mode frequency $\langle a^+a \rangle$ gives the number of detected photons in a bin and its time dependence gives the shape of the signal. The calculated signal has a characteristic non-exponential shape. Both shape and strength of the calculated signal depend on initial conditions and damping, which makes them situation and system specific. The time dependence of $\langle a^+a \rangle$ has the following simple structure:

$$n(t) = \langle a^+a \rangle = B_0 + \frac{B_1}{(1+\lambda_0 t)} + \frac{B_2}{(1+\lambda_0 t)^2}, \tag{1.6}$$

where B_i's (i =0,1,2) are positive coefficients. The coefficients depend on initial conditions and damping; the dependence is different for different coefficients. The value of B_0 is not large and can be much smaller than other coefficients. Any of the two coefficients responsible for time dependence can be much larger than the other, so that a biophoton signal may appear as decaying in a small interval. Further, the effective value of the exponent m of the decaying portion of any biophoton signal has to lie between 1 and 2. The initial conditions represent the response of a living system to light stimulation. The response depends upon many physiological and environmental factors and each of these factors influences the shape and strength of biophoton signal. The model does not distinguish between spontaneous and stimulated signals but describes them in a unified framework. The canonical parameters in the model are B_i's, which are to be used in characterising a biophoton signal and in investigating the influence of physiological and environmental factors on the signal. The investigations based on different parameters e.g. exponential decay constant and effective strengths, may lead to conflicting and erroneous results. These features of eq. (1.6) are unique and are not found in the classical solution of the model or in any model based on the conventional framework of photon emission. These are unusual features of a photon signal. All biophoton signals exhibit these unusual features, which makes us wonder the possibility of a quantum state of photon existing for a long time in living systems of macroscopic dimensions. It is, therefore, speculated that biophoton signal is a photon signal in a pure quantum state whose dynamics is described by the phenomenological model. The phenomenological model lacks a theoretical foundation and is silent about the mechanism, source and purpose of photon emission; it is a desperate though successful attempt to describe observed behaviour of biophoton signals. Establishing the quantum nature of biophoton signals and identifying their squeezed states will provide corroborative evidence of the validity of speculation and experimental foundation to the model.

2. PHOTO COUNT STATISTICS OF A SQUEEZED STATE

Evidence of the quantum nature of a photon signal comes from the measurement of conditional probability of no subsequent photon detection during a small interval Δ in the signal and the limiting value of the probability as average photon count $<n>$ in the interval goes to zero[7-9]. Any photon signal in a pure quantum state has the same limiting value that differs for a photon signal in the thermal equilibrium state. It is possible to measure this probability in a signal without measuring various other probabilities in 100ms for time intervals of a few tens of microseconds. Such measurements have been made in decaying and non-decaying portions of a few biophoton signals. The measured values at $<n> \approx 0.01$ are close to the limiting value of a quantum signal in biophoton signals and to the limiting value of a thermal source in photon signals of a light emitting diode (LED). The difference between measured values in photon signals emanating from living and non-living sources is observable only for smaller values of $<n>$. It is very small and unobservable for larger values of $<n>$.

The above evidence rests on two implicit assumptions – a biophoton signal remains in a pure quantum state during measurements and repeated measurements of photo counts are realisation of measurements made in the ensemble of systems specified by the quantum state. The repeated measurements determine the probabilities of detecting various numbers of photons in the quantum state if the ensemble remains unaltered during measurements i.e. only the phase of the state changes. The assumptions are of general nature and have been found to be valid in experiments involving laser signals. The ensemble associated with the quantum state representing a decaying signal changes with time, which makes the determination of probabilities questionable in decaying signals. The problem is partly avoided in measurements of durations much smaller than the typical decay time of the signal e.g. the measurements of 100ms duration in signals with characteristic decay time of $\approx$ 10s. The ensemble associated with the quantum state representing a non-decaying signal of spontaneous biophoton emission may be considered unchanging for much longer durations. The measurement of conditional probability in spontaneous biophoton signals are therefore, more reliable. It is emphasized that the limiting value of the conditional probability of no subsequent photon detection only establishes the quantum nature of a photon signal and the measurements of this probability do not give enough information to identify the quantum state. The identification of the quantum state requires knowledge of other remaining probabilities. Many probabilities are measurable in stable ensemble.

The probabilities of detecting different number of photons in a small interval or bin size are determined from a large number of measurements of photo count in successive bins. These measurements constitute a time series characterised by the bin size. The time series observed in biophoton signals are random and probabilistic. A time series is random if it does not show significant autocorrelation and is probabilistic if the occurrence of any specific value of photo count is unpredictable and its probability of occurrence approaches a definite value when the number of measurements in the time series becomes very large. A probabilistic time series determines the set of probabilities $\{P_{obs}(n); n=0,1,..\}$, where $P_{obs}(n)$ is the observed probability of detecting n photons in a bin. The random and probabilistic nature of a time series supports its identification to outcomes of successive measurements made in the ensemble representing the quantum state. The set $\{P_{obs}(n); n=0,1,..\}$ is called probability distribution of photo counts or photo count statistics. The set characterizes a spontaneous biophoton signal and is a means to identify the quantum state. The observed probabilities are essentially equal to the square of scalar products of signal and photon number states if complications arising from detection efficiency and geometry are ignored. There is an additional complication of the role of bin size. Many time series can be measured in a stable biophoton signal. The time series of different bin sizes yield different probability distributions. Since the scalar products of signal and photon number states are unaltered for an unchanging signal, there is a need to incorporate the effect of bin size. It has been incorporated through average signal strength measured with specific bin size.

The model predicts that the parameters specifying the squeezed state attain values independent of time in the non-decaying portion of a biophoton signal or in a spontaneous biophoton signal. The time dependence of the parameters is, therefore, dropped and the two complex parameters are expressed in the polar form as $\alpha=|\alpha|\exp(i\phi)$ and $\xi=r \exp(i\theta)$. The squeezed state describing a spontaneous biophoton signal is therefore, specified by four real parameters[10]. The calculated values of all measurable quantities in the squeezed state are expressed in terms of these four parameters. The calculated values of the probabilities $P_{cal}(n)$ of detecting n photons in the squeezed state are given by

$$P_{cal}(n) = |\langle n|\alpha,\xi\rangle|^2, \qquad (2.1)$$

with

$$\langle n|\alpha,\xi\rangle = \frac{1}{\sqrt{n!\cosh r}}\left[\frac{1}{2}\exp(i\theta)\tanh r\right]^{\frac{n}{2}} \exp\left[-\frac{1}{2}\left(|\alpha|^2 + \alpha^{*2}\exp(i\theta)\tanh r\right)\right] \times$$

$$\times H_n\left[\frac{\alpha+\alpha^*\exp(i\theta)\tanh r}{(2\exp(i\theta)\tanh r)^{\frac{1}{2}}}\right] \qquad (2.2)$$

where H_n is the Hermite polynomial of degree n. The calculated value of the signal strength k in the squeezed state for any bin size is given by

$$k = \langle a^+ a \rangle = \sinh^2 r + |\alpha|^2 . \tag{2.3}$$

The above calculations are valid for a single mode photon field and their use in a broadband biophoton signal implies the assumption of strong coupling of all photon modes. Since signal strength k is robust and well determined eq.(2.3) is used to express $|\alpha|$ as a function of k and r. It reduces the number of independent parameters of the squeezed state to three, namely r, θ, and φ.

3. MATERIALS AND METHOD

The measurements of time series were made in the spontaneous biophoton signals emitted by the dry and wet states of a sample of lichen species *Parmelia.tinctorum*. Lichen is suitable system for repeated measurements because of its very slow growth or decay[11-13] rate. A sample of lichen is a stable system and remains in the same metabolic state for many hours. A lichen sample has another advantage. Its metabolic activities change with its water content, so that the same sample exists in many metabolically different states. Two states are particularly significant- dry state and fully saturated wet states. The transition between dry and wet states of a sample is reversible and is affected by external means. A sample in the dry state, if made wet with some distilled water, switches to the wet state within a few minutes. The sample remains in the wet state for nearly 20h in normal laboratory conditions. It then starts becoming dry due to water evaporation. It attains the dry state in a few hours and remains in the dry state for weeks. The dry and wet states of the same sample emit different biophoton signals.

The measuring device has been described in many earlier publications. For the purpose of this experiment, the measuring device is essentially a small isolated sample chamber with a quartz window and a photo multiplier detector operating in single photon detection mode. Sample is placed in side the chamber in a quartz cuvette at a fixed position and the temperature of the sample chamber was maintained at 20^0C. A lichen sample in the dry state is put in the chamber and left as such for 15m; it eliminated the photons emitted in decaying signals. The sample usually starts emitting an almost constant flux of photons after about 5m. The emitted photons are detected in a time series in successive time bins of a fixed size. The number of measurements in a time series was 10,000. The bin size was then altered to measure another time series. The measurements of 13 time series with bin size equal to 50ms, 60 ms, 70ms, 80ms, 90ms, 100ms, 150ms, 200ms, 250ms, 300ms, 350ms, 400ms, and 450ms were made one after the other in this order. It took nearly 9h to complete these measurements. The sample was then made wet by pouring a small quantity of distilled water without disturbing the geometry. The sample in the wet state was left as such for another 30m. The spontaneous biophoton signal emitted by the wet state of the sample was detected in 14 time series. The order of measurements of first 13 time series was the same as in dry state and fourteenth time series was of bin size 500ms. Similar measurements were also made in a signal emitted by a light emitting diode (LED). LED is a non-living source of photons. Six time series each of 20,000 terms were measured one after the other with bin size of 5ms, 8ms, 10ms, 15ms, 17ms and 20ms.

Each time series is analysed for its statistical and quantum properties. The statistical properties are first four statistical moments namely Mean, Variance, Skewness and

Kurtosis and the quantum properties are Q-value (= Variance/Mean −1) and observed probabilities {$P_{obs}(n)$; n=0,1,..}. The statistical properties indicate the type of distribution. Non-zero Q-value indicates quantum nature of the signal. The parameters r, θ, and φ of the squeezed state are estimated by minimising the function $F_i(r, θ, φ)$ defined by

$$F_i(r, θ, φ) = \sum_n (P_{obs}(n) - P_{cal}(n))^2 \tag{3.1}$$

where summation is for all positive integer values of n and the subscript i identifies a time series. The function F(r, θ,φ) is the sum of squares of residual between observed and calculated probabilities. Only the values of n for which $P_{obs}(n)>0.001$ are included in the summation. The minimization programme is time consuming particularly in time series with k>15; the values of n in estimations involving these time series are restricted to first 15 terms. In the estimation of parameters common to various time series the minimization function is changed to F(r, θ,φ) given by

$$F(r, θ, φ) = \sum_i F_i(r, θ, φ) \tag{3.2}$$

where summation extends over time series for which common parameters are sought. The parameters common to two or more time series are estimated starting from the time series of 50ms bin size then including other time series one by one in the estimation in increasing bin size.

4. RESULTS AND DISCUSSIONS

The statistical properties of time series are similar to those observed earlier [1,14] in biophoton signals emitted by other living systems. The Mean value of a time series gives signal strength or k. The bin size of time series changes by an order of magnitude so that a graph of k versus bin size does not clearly bring out the deviations from the linear increase of k with bin size. In order to bring out the deviations average flux expressing signal strength as number of photo counts per 50ms is plotted for time series of different bin sizes in Figure1. The bin size is plotted in the X-axis and is in logarithmic scale. The average photon flux of the biophoton signal emitted in the dry state of the sample is nearly constant during measurements lasting for 9h. The average photon flux changes to a much higher value in the biophoton signal emitted in the wet state of the same sample. The value in the wet state is nearly five times of its value in the dry state. The average flux remains nearly unchanging during the first 3h and then increases slightly. The increase in the photon flux probably indicates an increase in the metabolic activities. Perhaps, the metabolic activities of the lichen sample take much longer time to stabilise after pouring of distilled water. The rate of change in the average photon flux is small; it can be ignored in any time series. Figure1 also gives Q-value of different time series of two biophoton signals emitted in the wet and dry states. Q-value is positive in all time series and lies in a small range. The non-zero value indicates non-classical nature of biophoton signals, the positive value indicates super Poisson distribution and small range indicates similar distribution of photo counts in different time series. The Q-value appears to be a characteristic of the signal; so is, perhaps, photo count distribution. Q-value in time series of dry state is higher than in time series of wet state. It indicates that photo count distribution in the dry state is more super Poissonian and there is some linkage between metabolic activities and Q-value or photo count distribution. The Skewness and Kurtosis, are non-zero in all time series; these moments lie in a larger range in different time series of a biophoton signal. It again points towards the non-classical nature of biophoton signals.

The parameters r, θ and φ of the squeezed state corresponding to a biophoton signal are estimated from photo count distribution of each time series. The estimated parameters provide an excellent representation of the observed photo count distribution of the time series. The estimated parameters differ only slightly in different time series of a biophoton signal. The estimated parameters appear independent of bin size and could be the stable properties of biophoton signals. The reason for the stability of parameters is not known. Perhaps, it is a consequence of quasi- stable nature of metabolic activities of the lichen sample. One wonders if all time series of a signal yield same parameters. The possibility was investigated by estimating parameters common with the time series of 50ms. The estimation was done using eq.(3.2) and number of time series included in the estimation was increased one by one. The parameters estimated from different number of time series are depicted in Figure 2.

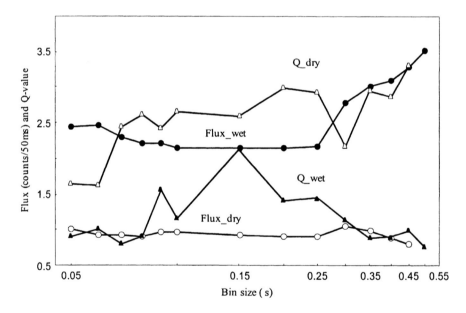

Figure 1. Photon flux and Q-value of time series of spontaneous biophoton signals: Photon flux in counts/50ms and Q value are depicted respectively by circles and triangles in time series of various bin sizes of two biophoton signals emitted by a sample of *Parmelia.tinctorum* in its dry and wet states. The depicting symbols are empty in the signal of dry state and are filled in the signal of wet state. The bin size is plotted along the X-axis in a logarithmic scale.

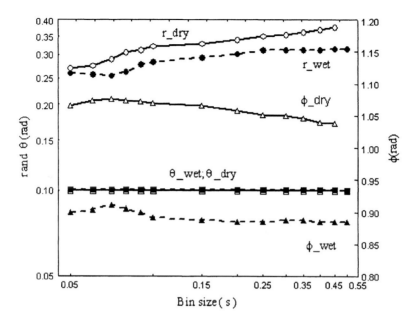

Figure 2. Estimated squeezed state parameters common to different number of time series: The parameters r,θ, and φ of biophoton signals emitted a sample of lichen species *Parmelia.tinctorum* are represented respectively by circles, rectangles and triangles. The symbols are empty for signal of dry state and filled for the signal of wet state. The right Y-axis gives the scale of φ. The scales of left Y and X axes are logarithmic. The parameters at any bin size are common to time series of lower bin sizes depicted in the figure.

The parameters depicted against a bin size in the figure are obtained by including all time series of lower bin sizes in the estimation. The parameters depicted at the largest bin size are same for all time series. These parameters specify the squeezed state of the signal. These parameters are new properties of the system. The figure shows that estimated parameters in time series of bin size <80ms deviate more from the final common values. This is an apparent effect arising from the peaked and unsymmetrical photo count distribution in these time series. The effect is considerably reduced in estimations weighted with higher power of observed probabilities. The value of F for parameters common to all time series is 0.0227 in the dry state and 0.00426 in the wet state. The number probability points included in the estimation of F is 392 in the dry state and 641 in the wet state. The higher strength of the signal and one more time series in the wet state are responsible for the difference in the number of points. The representation of the biophoton signal by a squeezed state is much better in case of the signal emitted by the wet state, perhaps, because the signal being more intense is much larger than the background noise. Since background is not subtracted from the measured values in a time series, its contribution to F is substantial in weak signals. Figure 2 also brings out the need for employing the parameters of the squeezed state in phenomenological investigations. The signal emitted by the wet state corresponds to smaller values of r and ϕ than the signal emitted by the dry state though θ has the same value in both signal. It means that the signals of wet and dry states differ in magnitude and phase of squeezing parameter. The enhanced metabolic activity of the wet state is reflected not only in the magnitude and shape of the decaying part but also in the strength and squeezing parameter of the non-decaying part of its biophoton signal. Biophoton signal provides multi parameter manifestation of metabolic activity. No appreciable change in the phase of the displacement parameter as a result of wetting and enhanced metabolic activity is an intriguing aspect of our analysis.

The above analysis was repeated in measurements made in a photon signal emitted by a non-living source LED. Six time series each of 20,000 terms were measured one after the other with bin size of 5ms, 8ms, 10ms, 15ms, 17ms and 20ms. The average of photo counts was 1.75, 2.83, 3.48, 5.24, 594 and 6.95 in these six time series. The average signal strength k increases linearly with bin size, indicating the emission of photons with nearly constant flux. The Q-value in these time series was respectively 0.040, 0.227, 0.106, 0.096, 0.160 and 0.094. The Q-value in the time series of this signal is much smaller than in the time series of biophoton signals. It could be considered as nearly zero, which indicates Poisson distribution of photo counts in these time series. Poisson distribution occurs in a signal in quantum coherent state or in a weak classical signal. The estimated parameters of the assumed squeezed state changes with bin size. The parameter r determining the magnitude of squeezing is the crucial parameter. Its estimated value determined from increasing number of time series is .092, 0.118, 0.128, 0.133, 0.134 and 0.135. The first value is obtained from the time series of 5ms, the second value from the time series of 5ms and 8ms, and so on. The value of r is small in these estimations and shows much larger variation compared to biophoton signals. The small value again indicates nearly Poisson distribution while larger variation suggests that the squeezed state description may not be valid.

The distinguishing features of time series in biophoton signals are high Q-value and stability of parameters specifying the assumed squeezed state[15]. Photo count distribution of a time series has a squeezed state description. The parameters specifying the squeezed state are easy to estimate in a time series with average counts per bin less than 10. The estimated parameters identify and measure fluctuations of the time series. The estimated parameters assume significance in a time series of high Q-value e.g. in time series of two spontaneous biophoton signals. The stability of the estimated parameters in different time series of a stable signal confirms their special significance. The parameters identify a structure in spontaneous biophoton signal hitherto considered structure less. The time series of different bin sizes yield same values of r, θ, ϕ, and $|\alpha|^2$/bin size in a biophoton signal. These are new properties of a biophoton signal and perhaps, of its emitting system. These properties will be system and situation specific like the properties of the decaying portion of a biophoton signal. The measurement of these properties construes a direct evidence of biophoton signal in a quantum squeezed state.

5. IMPLICATIONS OF SQUEEEZED STATE

The quantum nature of a biophoton signal and determination of its squeezed state have profound implications[16]. A biophoton signal in a pure quantum state implies the living system emitting the signal to be quantum, whose correct description only a quantum framework can provide. The semi-classical description of a living system as a composite structure of bio-molecules maintaining their separate identities is partial and incomplete. This description is successful only in explaining bio-molecule centric properties. It fails to describe the holistic properties of living systems. The holistic properties appear counter intuitive and non-local in character, whose description in the semi-classical framework requires correlation among biomolecules. The correlation required in describing some holistic properties can be introduced in a model by invoking some unknown and unspecific interaction but the bulk of holistic properties require correlation caused by interactions having superluminal communication. The needed interactions are arbitrary assumptions and missing links of the semi classical framework. Quantum framework does not face these problems; it describes a living system by a composite wave function having inherent correlation among biomolecules. The macroscopic duration and extension of biophoton signal requires the source of the signal to be a quantum system macroscopic in space and time. A model scenario of such a living system can be constructed in the semi classical framework though it will have a few arbitrary assumptions and missing links. The scenario will be an assertion of the quantum nature of "life" and will pull up the intimate connection between a living system and its biophoton signal. The intimate connection suggests the possibility of deciphering measurable indicators of vitality in a biophoton signal. The counter-intuitive and non-local features of a quantum system point out that quantum reality has components beyond classical visualisation. A glimpse of these components is provided by the parameters specifying the squeezed state of a biophoton signal. These parameters measure hitherto unknown properties of a living system. The unknown properties are new levels of reality of living systems to be explored. A brief discussion of the above three types of implications is given below.

5.1 Possible semi classical scenario of biophoton emission

The basic premise of the semi classical framework is biomolecules and their interaction. Biomolecules are irreducible building blocks of living systems and various properties of a living system including "life" and biophoton emission have to emanate from biomolecules. The mechanism for generating holistic nature in biophoton signals is a challenging problem, whose solution will unravel the mystery behind the creation and sustenance of life. The properties of biophoton signals provide a few pieces of information for unravelling the mystery. There is need to bind these pieces into picture. A rudimentary attempt is made in this direction with of desperate assumptions. The recent developments in information sciences offer some additional clues. The most important clue is provided by the properties of quantum selection that is more efficient than classical selection. A quantum entity can select the desired object from four possible objects in one step and from twenty possible objects in three steps in quantum selection[17]. The same entity selects the desired object from two objects in one step and from eight objects in three steps by employing classical selection. These properties could explain the basic facts of genetic code[18] namely, 4 bases of nucleotides, codons made up of three nucleotides and 20 amino acids. The explanation is based on the stipulation that the selections of the base pair by a nucleotide in replication, transcription and of the amino acid residue by a codon in protein synthesis are quantum selections. The quantum selection is the optimal selection strategy, which the living systems have learnt because of its evolutionary advantage. An essential requirement for quantum selection is the existence of entities and objects involved in the selection in pure quantum states. *It requires nucleotides (both DNA and RNA) to exist in pure quantum states at least during selections occurring in fundamental biological processes.* Many nucleotides may belong to a single quantum state. It is our contention that biophoton emission occurs in the transitions of the quantum states of nucleotides.

A quantum state of many nucleotides can be either decoupled or composite state of constituent nucleotides. The decoupled state maintains the identities of constituents and each nucleotide interact independently. In contrast, some constituent nucleotides loose their identities and act in a cooperative manner in a composite state. The decoupled state is no different from a classical state and the classical or semi classical framework can explain all its properties by invoking only local interaction of nucleotides. A composite state differs from a classical state and has three classes of properties: microscopic, local and holistic, and non local and holistic. The properties of the microscopic class are attributable to individual constituents and are identical to the properties in the decoupled state. The classical or semi classical framework can explain these properties. The properties of the local and holistic class are attributable to individual nucleotides but require correlations among nucleotides. The semi classical framework can explain these properties after invoking correlations causing interactions of unknown origin among nucleotides. The interactions are the missing link in the explanation. The properties of non-local and holistic class are manifestations of inherent non-locality of a quantum system. These properties have no classical counterpart and cannot be explained in the semi-classical framework. The detection of these properties requires special arrangement. The manifestation of any property involves transitions and energy transfers. Spontaneous transitions emit energy as photons. The class signature of a property is found in the properties of emitted photons. The transitions of the decoupled state correspond to the transitions of the states of individual nucleotides. Biochemistry investigates the energy transfers of these transitions. The transitions of composite states of nucleotides lead to biophoton emission and the emitted biophoton signals retain the holistic signatures of the composite states.

A biophoton emitting composite state does not need to be a state in which all nucleotides have lost their identities for all time. Only some nucleotides need to loose their identities to form a transient cluster (or clusters) and other nucleotides could remain decoupled for some time. The composite state probably contains many transient clusters of nucleotides. A cluster of nucleotides is a macroscopic quantum patch that could maintain its quantum compositeness only for a small duration. It is destabilised by the de-cohering interactions of the environment. It looses its quantum character or coherence and reverts back to the decoupled classical state after a small interval. The patch makes various quantum selections and stores the results of selections in a memory within this small interval. The scenario also needs a mechanism to activate a decoupled nucleotide and to form a quantum patch. The formation of quantum patch is opposed by the de-cohering interactions. The interplay of patch forming mechanism and de-cohering interactions ensures that many quantum patches of different sizes are continuously formed and destroyed. The nucleotides of a quantum patch need not be contiguous in the backbone. Even nucleotides distant in the backbone, may come close enough due to folding and on going dynamism to form a quantum patch. Further, nucleotides contained in a quantum patch at one time may belong to different patches at other time. The formation and destruction of patches continue through out the lifetime of a living system. The size and lifetime of a path and their distribution characterise the on going dynamism and a living system. It was mentioned earlier that the results of selections are stored in a memory for subsequent use. A living system should have a mechanism to safeguard the contents of the memory from destructions caused by de-cohering interactions. It could be achieved by a distributive memory in non-coding regions of nucleic acids in a manner similar to a neural network that allows retrievable of stored information even after a partial destruction of memory. Finally, repeated functioning requires a mechanism to reset the memory just before it is almost full. Various mechanisms invoked above are the missing links of the scenario. The scenario can be summarised by the following cycle of transitions: nucleotide in an inactive state → nucleotide in an active state → {formation of a patch} → {quantum selection} → {storage of the result of selection in a memory} → destruction of patch and transition of nucleotides to the inactive state → {resetting of memory}. The states within curly brackets are composite states of nucleotides. Each transition of the cycle will involve a characteristic amount of energy transfer in a specific mode. The various energy transfers in the cycle are situation and system specific components of the scenario. *It is our contention that normal biochemical machinery meets the energy needs of the cycle and energy is released in the form of biophotons.* The actualisation of the contention in various transitions of the cycle is elaborated below.

The scenario envisages a nucleotide to exist in two states of different energies. The lower energy state is called inactive and the higher energy state is called active. The decoupled classical state is made up of nucleotide in inactive states. The nucleotide in the inactive state preserves its identity in different interactions. All biochemical properties arise from the interactions of nucleotides in the inactive states. A nucleotide can make a transition from its inactive to active state by absorbing energy from the usual biochemical machinery of ATP-ADP cycle or its variant cycle. The transition to the active state confers two additional properties to the nucleotide: the nucleotide in the active state participate in quantum searches in case of need and tries to form a quantum patch with other neighbouring nucleotides in active states. All active nucleotides of a quantum patch participate in quantum searches simultaneously. Since a patch is made up of a large number of active nucleotides it contains a large amount of energy that is released in the transition of the patch to its decoupled classical state. The concept of quantum patch thus provides a mechanism to up convert biochemical energy in to biophotons. It is an essential ingredient of the scenario. The extent of energy up conversion depends on the size of a patch. The observed broadband spectral distribution of biophotons requires continuous variation in the size of quantum patches while emission mainly in the visible range requires the sizes of quantum patches to lie in a small range. The spectral range of biophotons is nearly the same in every living system. It means that quantum patches are of nearly identical sizes in living systems. Since the size of a patch depends upon patch forming and de-cohering interactions, the patches of nearly identical sizes is an assertion of the universality of nucleotides and nearly identical environment of nucleotides in living systems. The different sequences of nucleotides lead to different distributions of sizes. Similarly, biological processes change the environment to produce situation specific distributions of sizes. The distribution of sizes determines the coupling of different modes and spectral distribution of a biophoton signal. The scenario visualises in vivo folded and compressed nuclear material as an assembly of intermittent quantum patches of varying number of nucleotides continuously emitting biophotons in quantum transitions. The different patches of a single nucleic acid molecule are likely to emit photon signal of quantum nature but the quantum patches belonging to different cells in a living system also emit the photon signal of quantum nature. It is a vital missing link in the scenario that points out a connection between correlation among constituents of different cells and quantum coherence of biophoton. Perhaps, quantum patches do have a long-range interaction that binds quantum patches of different cells of a living system in to a single quantum state. The semi classical scenario needs this interaction but does not know its origin. Perhaps, it is the basis of life.

5.2 Vitality indicators and indices of vitality

The connection between life and quantum coherence of biophoton suggests the possibility of determining vitality indicators in biophoton signals. Some indicators may be able to grade living systems, food materials and vital fluids to act as vitality indices. The very presence of biophoton signal is an indicator of life, so that its properties can be considered as vitality indicators. The properties of its non-decaying portion are considered to be of limited utility because of ultra weak strength of this portion of the signal. The strength cannot be used to identify different systems and situations. The fluctuations that determine the parameters of the squeezed appear more promising. The promise will be ascertained in future phenomenological analyses. The decaying portion of a biophoton signal is very discriminating and has been put to use in many applications. Its two indicators have been identified, one indicator is quantitative and the other is qualitative. The quantitative indicator has the operational definition as number of counts in a definite interval i.e. $\int_{\varepsilon}^{\varepsilon+\Delta} n(t)dt$ in which ε and Δ are arbitrary positive constants determining the interval. These constants also determine the sensitivity of the indicator. Smaller values of ε and Δ give more sensitive indicator of vitality. There is a trade off between the sensitivity of the detector and its capability as vitality indicator. The number of counts in the first bin in any experimental set up is the most sensitive vitality indicator. The operational definition essentially measures a quantity related to the intensity of the signal. The operational definition of the qualitative indicator is the shape of a signal that is observable but not measurable. The shapes of biophoton signals can be compared to detect the difference. The qualitative indicator is more sensitive to various physiological and environmental factors than the quantitative indicator. The shape becomes measurable in a model. The measure of the shape in the quantum squeezed state model is the ratio of B_1 to B_2. The quantity B_1/B_2 gives the ratio of quantum correction to the classical contribution at t=0. The quantity can grade the shapes of various biophoton signals. It is very sensitive to changes in physiological and environmental factors and varies over a large range. We can define a related quantity by taking logarithm of the ratio B_1/B_2. The related quantity is the shape dependent vitality index VI_S given by

$$VI_S = \log\left(\frac{B_1}{B_2}\right). \tag{5.1}$$

Since the model ascribes vitality to quantum nature VI_S is a natural choice of vitality index. VI_S is very efficacious. The natural choice for the intensity related quantitative indicator in the model is VI_I given by

$$VI_I = B_0 + B_1 + B_2. \tag{5.2}$$

VI_I gives the intensity of the signal at t=0. Its value can be estimated from the analysis of a decaying biophoton signal and also from its extrapolation to t=0. Both VI_S and VI_I are situation and system specific. Each indicator is measurable and the measured value specifies relative vitality of the system. The indicators are, therefore, vitality indices of a living system. The index VI_S is also efficacious in identifying the germinating capacity of seeds and states of milk, vital fluids and other food materials.

5.3 New vistas of reality

The identification of biophoton signal as a photon signal in a quantum squeezed state radically transforms our perspectives of life and reality. A system endowed with life is a macroscopic quantum object that might perceive quantum reality. A quantum object behaves differently and quantum reality has subjective, intuitive and non-local features. The information about the quantum aspects of a living system is provided by biophoton signals. The information is in the form of parameters r, θ, ϕ, and $|\alpha|^2$/bin size. The four parameters are new characteristics of a living system and add four new dimensions of knowledge to fathom. The reality of a living system is much more deeper. This information has hitherto been hidden. The squeezed state description has opened up new vistas of reality.

The new information raises an interesting point connected with the possibility of many quantum states of a living system. All these states have the same non-living matter and classical disposition but they behave differently and emit different biophoton signals. The semi-classical framework does not know how to describe these states; it simply calls these states as moods of the living system. The mood is reflected in the biophoton signal, which offers a possibility of measuring the mood. Both mood and its biophoton reflection are situation and system specific. The parameters specifying a biophoton signal take continuous values, which gives a biophoton signal the capability to capture an immense diversity of the mood of a living system and its dependence on situations and systems. Investigations are needed to ascertain how diversity is reflected e.g. the results presented indicate that biophoton signals of dry and wet states of the sample have nearly same value of parameter ϕ. This is a new aspect of reality that needs exploring.

A squeezed state of photons is a minimum uncertainty state. It can transmit information coded in any or all of the four parameters in almost loss less manner with the speed of light. It is a new mode of communication that has remained hidden so far. The contained information is accessible from anywhere and could be preserved for posterity. Anyone with appropriate detecting mechanism can pick the information from anywhere; it will confer remote sensing capability in space and time. There is also a possibility of remote intervention in a suggestive mode[19]. The possibility hinges on the capability to identify a specific biophoton signal in a multitude of signals, to decipher its information content, to code a biophoton signal and to induce changes in other systems via biophotonic intervention. These are physical capabilities and any living system including human being could be in possession of these capabilities. A few living systems demonstrate these capabilities by generating detectable responses to specific biophoton stimuli e.g. onion roots, yeast cells and amphibian eggs[20]. The detection mechanism of these systems could be holistic and may not intensity based. It is conceivable that other living systems might be sensing some types of biophoton signals but may not be responding in ways known to us. It will be a case of lack of communication and not the lack of detection capability. The detecting capabilities can legitimise the concept of morphogenic field and its many variants.

6. REFERENCES

1. F.A. Popp, in: *Recent Advances in Biophoton Research and its Applications*, edited by F.A. Popp K.H. Li, and Q. Gu (World Scientific, Singapore,1992),pp. 1-46.
2. F.A. Popp, Properties of biophotons and their theoretical implications, *Ind. Jour.Exp.Bio.*41,391-402(2003).
3. R.P.Bajpai, in: *Biophotons*, edited by J.J. Chang, J. Fish, and F.A. Popp(Kluwer Academic,Netherland,1998), pp.323-338.
4. F.A.Popp and K. H.Li, in: *Recent Advances in Biophoton Research and its Applications*, edited by F.A. Popp K.H. Li, and Q. Gu (World Scientific, Singapore,1992),p. 47.
5. R.P.Bajpai, S. Kumar, and V.A. Sivadasan, Biophoton Emission in the evolution of a squeezed state of a frequency stable damped oscillator, *Applied Math and Comp.* 93, 277-288(1998).
6. H..P. Yuen, Two photon coherent states of the radiation field, *Phys. Rev.* **A13**, 2226(1976).
7. J. Perina, *Coherence of Light* (Van Nostrand Reinhold 1971),p.1.
8. R.P. Bajpai, Coherent Nature of Radiation emitted in Delayed Luminescence of Leaves, *Jour. Theo. Bio.*198 287-299(1999).
9. D. F.Walls, and G.J. Milburn, *Quantum Optics* (Springer Verlag, Heidelberg,1994), p.29
10. M. Orszag *Quantum Optics* (Springer Verlag Heidelberg,2000), 44
11. K.A. Kershaw, *Physiological Ecology of Lichens* (Cambridge University Press, Cambridge, 1985).
12. O.B.Blum in: *Water Relations in The Lichens* edited by V.Ahmadjian and M.E. Halle, (Academic Press, New York,1973) p..381.
13. P.K.Bajpai,R.P.Bajpai ,S.Chatterjee , A.Singh and G.P.Sinha in: *Biology of Lichens* edited by K.G.Mukerji, B.P.Chamola,D.K.Upreti, and R.K.Upadhyay (Aravali, New Delhi, 1999)pp.57-73.
14. M.Kobayashi and H.Inaba, *Applied Optics* 39 (2000), 183.
15. R.P. Bajpai, Biophotons of a Lichen Species *Parmelia.tinctorum*, *Ind. Jour. Exp. Bio.* 41 (2003),403-410.
16. R.P. Bajpai in: *Integrative Biophysics*,edited by F.A. Popp and L. Beloussov), (Kluwer Academic Publishers, Netherland, 2003)pp.439-465.
17. L.Grover Searching with Quantum Computers *Dr. Dobb's Journal*.April 2001. (also http://xxx.lanl.gov/abs/quant-ph/ **95122032** and Grover, L. K. "From Schrodinger's Equation to the Quantum Search Algorithm, http:// www.bell-labs.com/user/lkgrover/papers.html).
18. A. Patel Quantum Algorithms and the Genetic Code. http://xxx.lanl.gov/abs/quant-ph/ /0002037(2000).
19. Roy Ascott, *Reframing Consciousness*, (Intellect Books, London,1999).
20. L. Beloussov, A.B. Burlakov, and N.N.Louchinskaia, Biophotonic Patterns of Optical Interactions between Fish Eggs and Embryos, *Ind. Jour. Exp. Bio.* 41 (2003), 424-430.

Chapter 11

BIOPHOTONIC ANALYSIS OF SPONTANEOUS SELF-ORGANIZING OXIDATIVE PROCESSES IN AQUEOUS SYSTEMS

Vladimir L. Voeikov[1]

1. INTRODUCTION

Current concept of low-level photon emission (LLPE [*]) of biological objects considers it as chemiluminescence resulting from relaxation of electronically excited states generated in reactions with the participation of reactive oxygen species (ROS). In their turn oxygen free radical metabolites and other ROS had been considered until recently either as by-products of "normal" biochemical processes or at best as exotic chemical products of specialized immune cells designed to destroy alien viruses and bacteria. Thus, inasmuch as processes with ROS participation are still regarded by the majority of bio-medical scientists as auxiliary to "normal" biochemistry, LLPE which accompany these processes is looked upon as irrelevant to the performing vital functions.

Another point of view on LLPE from living matter is based on the notion that it originates from a delocalized coherent electromagnetic field that is tightly coupled to metabolic processes. In this context LLPE is termed as "biophotonic emission" and coherence theory "assigns to the presumably phase locked and mode coupled photons from DNA a permanent regulatory activity within cells and also between cells"[1]. However, this theory does not specify the source of energy which continuously pumps the biophotonic field.

At the same time it should be reminded about the works of Alexander Gurwitsch, the pioneer of the field of biophoton research. LLPE discovered by him more than 80 years ago was named "Mitogenetic radiation" because it performs a major biological function – it stimulates mitoses in competent cells. Many features of this radiation indicated its coherent nature (in a wide sense of this term), and at the same time it was proved by Gurwitsch that oxygen-dependent free radical reactions were indispensable for its

---

[1] Vladimir L. Voeikov, Faculty of Biology, Lomonosov Moscow State University, Moscow, 119234, Russia
[*] Abbreviations: Low level photon emission – LLPE, reactive oxygen species – ROS, electron excitation energy – EEE, electronically excited states – EES, chemiluminescence – CL, Maillard reaction – MR.

origination (see[2] for the details and references therein). Gurwitsch's discoveries and ideas had initially attracted much interest and initiated in 1930 wide studies of mitogenetic radiation. Regrettably, they are nearly completely forgotten because of an unfortunate combination of complex historical and psychological factors.

In recent years the attitude to oxygen radicals as to only hazardous by-products of "normal" metabolism started to change. More and more evidence is accumulating that they play an important if not a central role in all aspects of regulation of biological functions[3]. Still, the connection of their bioregulatory role with biophotonic field is not yet recognized. Here we'll present arguments approving such a connection and illustrate them with evidence in favor of intrinsic property of the processes running in aqueous milieu of living systems with ROS participation to self-organization which by itself is a crucial property of living matter.

2. PATHWAYS OF OXYGEN CONSUMPTION IN LIVING ORGANISMS

Practically all living organisms on Earth gain energy for performing their functions from oxygen-dependent oxidation of various "fuels". It is generally accepted that in the course of aerobic respiration oxygen is used in the terminal stage of the chain of oxidative phosphorylation in mitochondria. It is reduced there by cytochrome c reductase which transfers 4 electrons (together with 4 protons) to oxygen molecule. Oxygen plays here the role of the "trash box" for electrons that had already used their redox potential for the synthesis of ATP molecules while traveling along the respiratory chain. High initial energy potential of electrons obtained from food dehydrogenation is coined in this way into multiple small portions: energy released in hydrolysis of 1 ATP molecule does not exceed 0,5 eV, equivalent to IR-photons with $\lambda \sim 2,5$ μ. Oxidative process in which energy is received in such a form, in principle, in a form of heat, is analogous to putrefaction or smoldering. This pathway of oxygen utilization for gaining energy in the form of ATP is considered to be practically the only one in "normal" biochemistry.

But smoldering is not the only way of energy gaining from oxygen-dependent oxidative processes. Another one is burning (combustion). The difference between the two is that in the latter O_2 is directly reduced with electrons (which often go together with protons). It should be reminded that oxygen is a unique molecule, which in its ground state contains 2 unpaired electrons, and at the same time each single oxygen atom contains also two unpaired electrons. Thus a total of 4 electrons are needed to equilibrate all its unpaired electrons. "Heavy" energy quanta are released at each act of electron pairing. On the way of one-electron O_2 reduction in few discrete steps a total of 8 eV may be received as O_2 receives electrons sequentially according to the following scheme:

1) $^{\bullet}O_2^{\bullet} + e^- (+H^+) \rightarrow HO_2^{\bullet}$; 2) $HO_2^{\bullet} + e^- (+H^+) \rightarrow H_2O_2$;

3) $H_2O_2 + e^- (+H^+) \rightarrow H_2O + HO^{\bullet}$; 4) $HO^{\bullet} + e^- (+H^+) \rightarrow H_2O$

Intermediate products of its reduction are either free radicals ("•" symbolizes an unpaired electron), or metastable molecules such as H_2O_2 possessing high chemical potential. All of them are defined as Reactive Oxygen Species (ROS).

One-electron oxygen reduction leading to the emergence of ROS is still considered by many apprentice and even connoisseurs in bio-medical research, as a dark side of dioxygen biochemistry[4]. Such an attitude stems from innumerous studies on mostly *in vitro* biochemical and cell culture models, where it has been demonstrated that ROS easily damage lipids, proteins and nucleic acids. This damage may be a massive one, because a single free radical may in principle initiate chain reaction in the course of which many molecules may be affected. However, on this way not only very active particles originate, but on each step a quantum of energy is released that belongs to the range of EEE rather than to the range of energy of vibrational and rotational excitation as in smoldering. Thus, one-electron oxygen reduction is genuine burning, which may be accompanied by release of light photons.

Nearly anonymous attitude to ROS as universal pathogens is now gradually eroding. An exponential growth of the number of works devoted to the bio-regulatory role of ROS (see refs. in[3, 5]) is observed now. Adequate reactions of cells upon hormones, neurotransmitters, cytokines, upon physical stimuli (light, temperature, mechanical stimulation) finally depend on ROS "background". ROS added to cultured cells may induce in them normal physiological reactions, from reversible activation or inhibition of certain enzymatic chains to switching genome activity. However, until now no consistent model of the mechanism of regulatory ROS action is suggested.

But how much oxygen consumed by animals goes to ROS generation? Until now in numerous papers reference is made to estimates that were obtained decades ago in *in vitro* systems, e. g. in isolated mitochondria that were for a long time considered to be the major source of ROS due to accidental "escape" of electrons from electron-transport chain directly to oxygen. According to these estimates the share does not exceed one, at most – a few percent. However, recent direct measurements had shown that this share may be remarkably high. For example, in cleavage stage chicken embryos 70% of O_2 consumed is one-electronically reduced, and this share decreases to 30% at the blastocyst stage. This is due to increase of mitochondrial O_2 consumption, while absolute quantity of directly reduced O_2 does not change[6]. In isolated rat aorta more than 27% of all O_2 taken by aorta tissue is converted to superoxide anion $(O_2^{\bullet-})$[7].

Actually, these results are not unpredicted. There are plenty of ways for one-electron oxygen reduction in an organism. For example, NADPH-oxidase, the enzyme that directly reduces O_2 to $O_2^{\bullet-}$ that was initially considered to be specific for immune cells such as neutrophils and eosinophils, is now found in practically all types of cells of multicellular and in unicellular organisms[5]. It is one of the group of other enzymes that are able reduce O_2 while performing their normal functions. ROS are also generated in multiple non-enzymatic reactions that continuously proceed in any organism, such as the amino-carbonyl (Maillard) reaction[8].

On the other hand stationary levels of ROS in cells and tissues are very low (of the order of 10^{-6} M or less) due to the universal presence in cells and tissues of the so-called antioxidative system represented by a variety of enzymes and low-molecular weight "antioxidants". For example, the ubiquitous enzyme, superoxide dismutase, nearly immediately converts $O_2^{\bullet-}$ to H_2O_2 and O_2. H_2O_2 is in turn very quickly degraded by catalase. The question of why an organism directs such a huge share of valuable oxygen for ROS production and eliminates them practically immediately is not discussed in the literature.

3. BIO-REGULATORY FUNCTIONS OF PROCESSES ACCOMPANIED WITH GENERATION OF ELECTRON EXCITATION ENERGY

A flaw in the efforts to explain the mechanism of bio-regulatory action of ROS by analogy with molecular signaling mechanisms may be related to disregard of the unique high energy output of reactions with ROS participation. It is known that when unpaired electrons recombine energy quanta sufficient to induce electronically excited states (EES) of reaction products are liberated. Such reactions are usually accompanied with chemiluminescence (CL) and other forms of luminescence. A probable role of EEE, of EES of biomolecules, of LLPE has received little attention in contemporary cell physiology. This originates possibly from underestimation of the quantity of oxygen that is directly reduced by living organisms in the course of their normal functioning. Thus, we decided to study carefully if oxidative processes with ROS participation accompanied with PE, indicating of EEE generation, may play a regulatory role in a biological system such as blood and also in aqueous model systems in which such processes develop spontaneously.

We have chosen human blood as an experimental system not only because it donates oxygen for all cells and tissues of an organism, but because it itself contains consumers of oxygen. It has been shown recently that neutrophils and eosinophiles representing the majority of leucocytes, even in their resting state convert all oxygen that they consume into free radicals[9], not to mention the production of ROS in their stimulated state when it increases more than 10-fold. Besides a short time ago it was discovered that immunoglobulins catalyze direct oxygenation of water with the appearance of H_2O_2 and even ozone[10].

Blood is a complex biological system, but at the same time it is an aqueous system. ROS that are initially produced in it represent hydrophilic particles and the processes in which they participate may proceed in aqueous phase. For revealing basic properties of such processes we used a model aqueous system – solutions of amino acids in which slow oxygen-dependent oxidative processes spontaneously develop.

As the reactions with ROS participation are accompanied with LLPE, the principle method that we used was monitoring PE from experimental systems under different conditions with single photon detectors predominantly for prolonged periods of time to reveal dynamic peculiarities of these processes.

3.1. Biophoton Emission from Non-diluted Human Blood.

It can be seen from Fig. 1 that addition to whole non-dilute human blood of zymosan – a specific inducer of respiratory burst of neutrophils, results in a significant elevation of LLPE from blood that lasts for many hours (left Y-axis). In the presence of luminol, a fluorescent probe for reactions with H_2O_2 and some other ROS participation, maximal intensity of PE elevates about 100-fold (right Y-axis), while its kinetics in the first approximation remains the same. Thus very high optical density of blood does not interfere with the registration of PE from it.

We've also found that PE may be registered even in "resting" blood after addition of lucigenin (a luminescent probe for superoxide anion). However, lucigenin-dependent PE decreases with dilution of blood with physiological salt solution, while luminol-dependent PE in blood in which respiratory burst has been induced is not affected by such dilution[11]. Hence, ROS generation in blood accompanied with generation of EEE goes on persistently even under non-stimulated conditions, provided by a close interaction of oxygen donors (erythrocytes) and oxygen acceptors (neutrophils). Under such conditions lucigenin-dependent PE may be observed even from small portions of blood for many hours.

The very fact that pronounced PE may be registered from non-diluted blood -- a highly opaque liquid because of very high concentration of hemoglobin – indicates that hemoglobin packed in erythrocytes does not quench efficiently PE. However, if free hemoglobin is added to blood at a concentration of only 0,5% of the amount present, lucigenin-dependent PE practically disappears[12]. Taking into account that concentration of hemoglobin in erythrocytes may reach a value as high as 35-40% (hemoglobin can not reach such high concentration in a free solution) one may suggest that hemoglobin in erythrocytes is present in a liquid crystalline state. In such a form it may provide transfer of excitation energy over long distances without its dissipation, unlike hemoglobin in a solution that absorbs and dissipates energy of electron excitation.

Under certain conditions PE from blood or suspensions of isolated neutrophils may gain oscillatory patterns in which amplitude of oscillations may reach up to 25% of the mean PE intensity (Fig. 2). These conditions include lack of agitation of a sample, an optimal buffered medium containing nutrients (in the case of neutrophil suspensions) and access to air.

Figure 2 illustrates prominent oscillations of luminol-dependent PE from neutrophil suspensions that develop about 1 hour after the initiation of respiratory burst with zymosan. It can be seen that they are much more prolonged and prominent when the suspension is in contact with air. In the absence of air oscillatory behavior can also be seen, however, PE fades much earlier and amplitudes of oscillations are smaller.

Does EEE generated in blood in the reactions with ROS participation and expressed as LLPE have any functional role for the processes taking place or it is just a by-product of these processes? The answer to this question was obtained in experiments illustrated in Figure 3. If a test-tube with blood in placed in a glass vial, its walls reflect some part of the irradiated photon flux back into the blood.

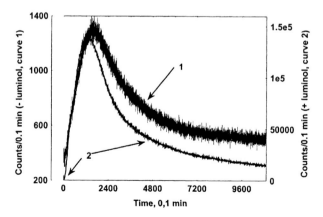

Figure 1. Photon emission from non-diluted blood (0.1 ml) without luminol (curve 1, left Y-scale) or with it (curve 2, right Y-scale) after induction of respiratory burst by zymosan (without zymosan and luminol PE from blood sample was 250-300 counts/6 sec).

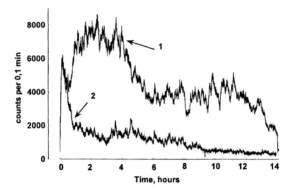

Figure 2. Sustained oscillations of luminol-dependent PE in neutrophil suspensions (20000 cells in 0.1 ml of M-PRM medium; during respiratory burst induced with zymosan. Curve 1 – aerobic conditions, Curve 2 – suspension isolated from air.

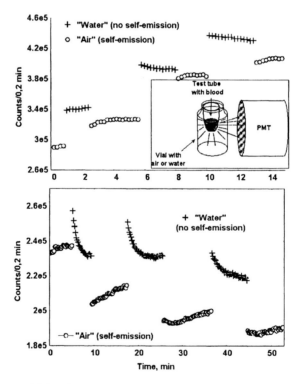

Figure 3. Effect of partial light reflection from the inner surface of the glass vial upon luminol-dependent PE at different stages of respiratory burst in whole blood. Up panel – 1 hr after zymosan addition, PE is approaching maximal intensity (insert – scheme of experimental setup). Lower panel – 20 hrs after respiratory burst initiation, respiratory burst decay.

BIOPHOTONIC ANALYSIS OF OXIDATIVE PROCESSES

If a vial is filled with water, back reflection of photons is virtually absent. In fact, transfer of a test tube with blood from an empty to a water-filled vial results in a sharp leap of PE resulting from increased escape of photons from the system due to the immersion effect. Besides this one can notice that self-irradiation or absence of it results in different dynamic patterns of PE from blood. At a stage when PE was approaching maximal values, blood transfer to the water-filled vial after a period of its presence in air-filled vial (period of self-irradiation) results in a significantly higher level of PE that could be expected without its self-irradiation (Fig. 3, left).

Especially prominent is the effect of self- irradiation at the stage of PE decay (Fig. 3, right). Without self-irradiation PE rapidly decays, while when this blood is again self-

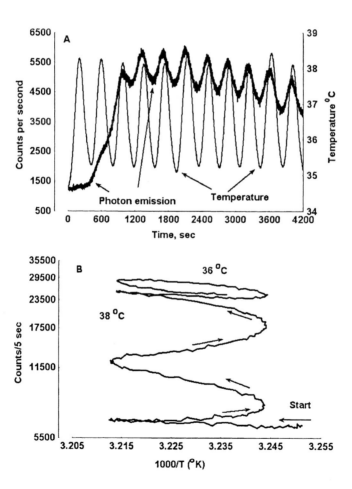

Figure 4. Temperature dependence of luminol-amplified PE from blood (3 ml) in which respiratory burst is induced with zymosan. (A) Original data. (B) Ahrrenius plot for the stage of development of respiratory burst (ca. up to 1700 sec).

irradiated, respiratory burst starts to "kindle" again. It should be recognized that during self-irradiation blood receives no more than few thousands photons/sec, equivalent to heat energy gain of about $10^{-17} - 10^{-18}$ cal/sec. Thus, functional significance of ultra-weak photon fluxes can not result from gaining additional energy by blood. It is likely that EEE plays a signaling function in systems that are very far from energy equilibrium, being at the same time at thermal equilibrium with their surroundings. In a certain sense such a system as blood may be looked upon as a physical active medium.

This conclusion is supported by the results of studies of the dependence of PE from blood upon variations of its temperature. Blood sample in which respiratory burst was initiated with zymosan in the presence of luminol was sequentially heated and cooled in the range of physiological temperatures (Fig. 4). As PE is the integral result of complex chemical reactions one might expect that its intensity should change in accord with temperature variations in blood.

Actual results indicate that this is not the case. PE rate practically does not depend on temperature at the stage of respiratory burst development. One can see even the negative slope for activation energy on the Ahrrenius plot (Fig. 4, B) for this stage. That suggests that blood is such an intense energy source by itself that it can easily overcome temperature energy losses when it is cooled. Only when PE intensity reaches maximal values temperature dependence shows itself, though a significant hysteresis is observed (upper part of the Ahrrenius plot) indicating that blood continues to oppose energy losses. Such dynamic properties of blood indicate that PE from it results from processes comprising a web of reactions with multiple positive and negative feedbacks. High level of coupling in these processes follows also from oscillatory behavior of PE from blood and neutrophil suspensions (see Fig. 2). In general, all these results argue that significant part of internal energy that provides blood non-equilibricity is represented by EEE

3.2. Autoregulation in Model AqueousSystems Related to ROS Production and EEE Generation.

Many features of the processes with ROS participation giving rise to EEE generation in blood may be considered to be related to its complex material composition. However, blood is essentially an aqueous system, so it is interesting to distinguish common features of oxygen-dependent oxidation reactions accompanied with PE going on in blood and in much more simple aqueous systems. To perform this comparison we used aqueous solutions of amino acids in which luminescent reactions emerge after addition of H_2O_2 or brief irradiation with an ultra-weak source of UV-light, and solutions containing amino acids and active reducing carbonyl compounds such as glucose or methylglyoxal[13]. Analysis of processes taking place in model systems in fact revealed some fundamental similarities between them and those observed in human blood.

We have previously shown[13], that addition of an aliquot of concentrated H_2O_2 to aqueous solutions of different amino acids to the final H_2O_2 concentrations of 0,1-0,4 M initiates in this solution the reaction of oxidative deamination of an amino acid that is accompanied by PE in the presence of different fluorescent compounds. The process is characterized with a definite lag-period after which a stage of acceleration of PE rate follows and then PE slowly declines. PE rate very well correlates with the rate of accumulation of ammonia at low (micromolar) concentrations of fluorophores, but at high concentrations of the latter both PE growth and deamination of amino acid are retarded. Thus, PE rate reflects the rate of the major reaction going on in the system.

Dependence of the rates of both parameters on concentration of fluorophores indicates that the chemical process is coupled to generation of EEE in the system[14].

One of the peculiarities of oxidative processes in model systems was the necessity to overcome some threshold conditions for these processes to develop. For example, PE does not develop in a solution of H_2O_2/asparagines if the concentration of the latter is below 25 mM, but if its concentration exceeds 30 mM the wave of emission under the same condition always develops. Thresholds (critical phenomena) are an intrinsic property of runaway chain reactions[13, 14]. Existence of critical phenomena in the studied processes indicate that they belong to a family of chain reactions with delayed branching.

Another specific feature of branching chain reactions is a significant deviation of their kinetic characteristics from the classical law of Ahrrenius. In studies of temperature dependence of PE from solutions of amino acids performed in the same manner as with blood (cf. Fig. 4) similar hysteresis behavior was observed, though in blood hysteresis was more pronounced especially at the stage of PE development.

The second model system studied by us is the reaction that develops in aqueous solutions of active carbonyls (such as methylglyoxal) and amino acids, polyamines or even such simple amines as ethylamine. This model system is more complex, yet it is more physiological. The reaction that develops in it is known as amino-carbonyl or Maillard reaction (MR) and it continuously proceeds in a non-enzymatic way in the internal medium of living organisms. Disturbances in MR are considered to play a significant role in the pathogenesis of diabetes and other metabolic disorders.

Though MR is thoroughly investigated for many decades, its mechanism is far from being clear. Probably its complexity is related to the fact that already at the first stage of the reaction – the formation of Schiff bases, there appear conditions for one-electron oxygen reduction, for ROS and other free radicals generation[15], and, hence, for the emergence EEE. Owing to the appearance of fluorescent and colored products in the course of MR, propagation and modulation of EEE may occur, seriously affecting the course of the process. However, this feature of MR practically does not currently attract attention.

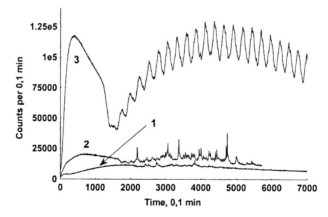

Figure 5. Dependence of PE patterns from MR developing in a solution of methylglyoxal (10 mM) on ethanolamine (EA) concentration. 1 - 5 mM EA, 2 - 10 mM EA, 3 -30 mM EA. Other conditions: total volume – 20 ml, surface area of the reaction mixture – 400 mm^2, temperature – 20 °C, pH 10.3.

Monitoring PE patterns from MR of different chemical composition under non-stirring conditions for prolonged periods of time allowed to reveal interesting regularities dependent on the generation of EEE that weakly depended on particular amino- and carbonyl reagents taken initially[16, 17]. One of the major results of this study is defining the range of conditions under which regular and irregular oscillations of PE with prominent amplitude develop that last for many hours or even days. Thus for the first time it was shown that the process of self-organization may develop in such systems under conditions close to physiological. Development of oscillations depends on many factors including threshold concentrations of reagents, volume to surface ratio of the reaction system, temperature range, etc.

For example, the PE pattern of the reaction developing in a solution of methylglyoxal and ethanolamine is critically dependent on ethanolamine concentration (Fig.5).Oscillations are practically absent at 5 mM amine concentration, they are small and early disappearing at 10 mM, but when the concentration is 30 mM very prominent oscillations are observed that last for many hours without damping. Noticeable that the integral number of photons emitted in the latter case, reflecting the intensity and yield of chemical processes, is orders of magnitude larger in comparison to the formers, though the concentration of the amine is only 3-fold larger.

Analogous threshold concentration for methylglyoxal in the presence of fixed glycin concentration was found. Under all fixed conditions oscillations practically did not develop at 2 mM of methylglyoxal and were prominent with simultaneous strong elevation of the mean PE intensity at 4 mM of methylglyoxal.

Oxygen plays a critical role in the development of PE from MR. The first wave of PE observed in Fig. 5 is related to consumption during the reaction of oxygen initially dissolved in water, and the second one is due to oxygen that diffuses to the reaction system from ambient air. If there is no contact of the reaction system with the air, the second wave of PE on which oscillations emerge does not develop for many hours. One could expect that the larger is the surface area, the better the reaction develops.

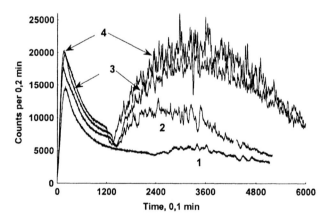

Figure 6. Dependence of PE patterns from MR developing in a solution of methylglyoxal (10 mM)/ethanolamine (30 mM) on surface area of the reaction system. 1 – 78 mm^2, 2 – 63 mm^2, 3 – 38 mm^2, 4 – 9 mm^2 at 4 ml of a solution in carbonate buffer (50 mM, pH 10.3).

Paradoxically, it turned out that excessive aeration of the reaction system prevents the second wave development and the emergence of oscillations. Figure 6 illustrates that at constant volume the less is the surface area, the more prominent is the second wave of PE, the longer it sustains, and the larger are the amplitudes of PE intensity oscillations. Here also a strong non-linear dependence of the parameters of PE on surface area may be noted. Taking into consideration that the intensity of PE reflects intensity of oxidative processes in the course of which EEE is generated, these results indicate that optimal conditions for the development of these processes imply certain restriction of oxygen diffusion to the reaction system.

Why does it happen? Our studies have shown that the processes that are accompanied with PE modulated with oscillations proceed in the uppermost part of the reaction system, in the vicinity of the water/air boundary. Besides, the process starts to develop only when all the oxygen initially dissolved in water is consumed and the latter gains a high reducing (negative) potential[16].

This can be seen in Figure 7, where the results of simultaneous measurements of PE and redox potential changes in the uppermost part of the reaction system are presented. Note also, that after the second wave of PE starts to develop, mean value of redox potential also starts to increase and it also oscillates with exactly the same periodicity as PE (this was confirmed by the Fourier analysis of the respective time series). Thus, practically full correlation between the two processes exists.

Measurement of redox potential changes in lower parts of the reaction system did not already reveal such a correlation, but the data suggested the appearance of complex spatial patterns of energy distribution along the reaction system (from its top to the bottom). For example, the redox electrode imbedded in a middle part of the reaction

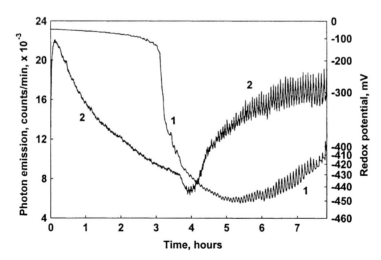

Figure 7. Simultaneous measurement of redox potential near water/air interface (1, right Y-scale) and PE (2, left Y-scale) from MR developing in a solution of methylglyoxal (10 mM)/ethanolamine (30 mM), 40 ml, surface area 440 mm², carbonate buffer (50 mM), pH 10.3.

vessel initially registered a very significant (down to -0,5 V) decline in the redox potential. Subsequently, very long period waves (wave periods reach several hours) of the redox potential with amplitudes reaching 0,3 V are registered. Thus intense oscillations of energy potentials apply to the whole reaction system, but their frequencies are different in different parts indicating that in the course of the whole process highly non-linear waves of energy continuously travel in the reaction system.

Among other peculiarities revealed in the course of studies of energy processes characteristic for MR there is one more that is common to blood. It was mentioned above that self-irradiation of blood may reactivate respiratory burst in it especially at the stage of PE decline. In the case of MR we found that when two cuvettes with faded MR were placed together before a photomultiplier, integral PE was significantly higher than could be expected from simple addition. After such a contact PE from both reaction systems was also for some time higher than before.

3.3. Probable role of water.

Water is the major chemical species in the systems studied here as well as in all biological systems. Formally, its concentration is 2-5 orders of magnitude higher than that of any other individual substance present in it. It may play a significant though yet unrecognized role in the processes in which ROS are produced and EEE is generated in aqueous systems. In fact, when we studied oxidative processes induced by H_2O_2 in aqueous solutions of amino acids unexpected elevation of H_2O_2 concentration at the stage of exponential PE rate acceleration was observed. In some experiments this elevation reached 10-12% of the initial H_2O_2 concentration[14]. The only possible source for new H_2O_2 in this reaction system is direct oxidation of water with oxygen. This reaction seemed to be very improbable under mild conditions of our experiments, but it now turned out that it can indeed occur. First, it was recently discovered that all immunoglobulins irrespective of their class and specificity efficiently catalyze oxidation of water with singlet oxygen[10]. Appearance of plenty of electronically excited singlet oxygen in our reactions systems, where EEE is continuously released is inevitable. Spatial and temporal self-organization which is essential for any forms of catalysis also takes place here. Thus, the conditions for water oxidation by oxygen may exist.

On the other hand it has been recently demonstrated by several groups of authors that overall water splitting may take place under very mild conditions[18,19]. In the course of this process ROS and other free radicals such as hydrogen atoms arise. It is interesting to speculate that it is due to active involvement of water that strong oscillations of redox potential may appear reaching amplitudes of hundreds of millivolts, as we observed in the course of the development of MR. Taking into consideration that concentrations of organic reagents (carbonyl and amino compounds) do not exceed tens of millimolar, and that standard redox potentials between their reduced and oxidized forms are very modest, such strong variations of redox potential may be explained by the appearance in water of such active reducers and oxidizers such as hydrogen and oxygen in relatively high concentrations and the possibility of their spatial, though temporal separation in these seemingly homogenous reaction systems. Only new experiments and experimental approaches may refute or support these speculations.

4. GENERAL CONCLUSIONS

We demonstrated that processes in which ROS are produced and EEE is generated continuously in blood as well as in model aqueous systems. EEE arising in these processes influences by a feedback mechanism ROS production.

A lot of common features in the behavior of such processes proceeding in blood and model aqueous systems can be noted. This indicates the possible identity of the most basic mechanisms for realization of regulatory action of the ROS through generation of EEE in aqueous system, independent of the level of their complexity.

At this stage it should be pointed to the conditions for the development of oscillatory regimes in MR, because they may be important for understanding ubiquitous regulatory functions of ROS in biological systems. These conditions include:

1. The existence of a sharp gradient of oxygen between the reaction system consuming it and the gas phase which supplies it;
2. Restriction of rate of O_2 supply to the reaction system;
3. Elevation of concentrations of strong reducing compounds in the reaction system over a certain threshold value;
4. Existence of a mechanism allowing the acceleration of oxidation processes as soon as they commence, resulting in a fast decrease of O_2 concentration below the threshold level.

The same conditions of oxygen consumption are characteristic of any aerobic living cell especially under the condition of high activity when it is actively breathing to make a lot of energy. Under normal physiological conditions oxygen supply is restricted by its limited solubility in water (for free living cells) and by special regulatory mechanisms in multicellular organisms. In periods when in the nearest cell vicinity oxygen is exhausted, the concentration of reducing equivalents in a cell increases to a threshold when newly coming oxygen starts a new inflammation. In the course of this inflammation EEE is generated and it serves as energy of activation for faster burning that results in oxygen depletion. Thus spontaneous oscillatory regimes of cell breathing develop, and these oscillations may play a role of pacemakers of all vital processes in cells and their communities. In fact, such oscillatory "breathing" – oxygen consumption – has been demonstrated in single beta-cells, stimulated by the addition of glucose[20].

EEE, emerging in the considered processes in blood and model systems may be functionally significant (participate in the orderliness of the considered systems) if the following conditions apply:

1. Absorption of EEE by the components of the system shifts them into a functionally active state;
2. A system should possess own mechanisms for EEE generation;
3. A system may amplify, concentrate and prevent free dissipation of EEE;
4. EEE generation in time and space is ordered.

All these conditions, as it has been demonstrated here, more or less prominently realize in all the studied systems, including blood, that ubiquitous for living systems processes with ROS participation may regulate a wide spectrum of biochemical and physiological functions. No less important is that if such processes commence in initially disordered systems they have the intrinsic property for self-organization under rather wide boundary conditions.

5. REFERENCES

1. F.-A. Popp, in: *Recent Advances in Biophoton Research and its Applicaions*, edited by F.-A. Popp, K.H. Li, and Q.Gu (World Scientific, Singapore, New Jersey, London, Hong Kong, 1992), pp. 1-46
2. V.L. Voeikov, in: *Integrative Biophysics. Biophotonics.* edited by F.-A. Popp and L.V. Beloussov (Kluwer Academic Publishers, Dortrecht, The Netherlands, 2003), pp. 331-360.
3. V.L. Voeikov, Reactive oxygen species, water, photons, and life, *Rivista di Biologia/Biology Forum* **94**(1), 193-214 (2001)
4. J.S. Valentine, D.L. Wertz, T.J. Lyons, L.-L. Liou, J.J. Goto, and E.B. Gralla, The dark side of dioxygen biochemistry, *Current Opinion in Chemical Biology*, **2**(2), 253–262. (1998)
5. V.L. Voeikov, in: *Biophotonics and Coherent Systems. Proceedings of the 2nd Alexander Gurwitsch Conference and Additional Contributions.* edited by L. Beloussov, F.-A. Popp, V. Voeikov, and R. Van Wijk (Moscow University Press, Moscow, 2000), pp. 203-228.
6. J.R. Trimarchi, L. Liu, D.M. Porterfield, P.J.S. Smith, and D.L. Keefe, Oxidative phosphorylation-dependent and -independent oxygen consumption by individual preimplantation mouse embryos, *Biol. Reproduct.,* **62**(6), 1866–1874 (2000)
7. H.P. Souza, X. Liu, A. Samouilov, P. Kuppusamy, F.R.M. Laurindo, and J.L. Zweier, Quantitation of superoxide generation and substrate utilization by vascular NAD(P)H oxidase, *Am. J. Physiol. Heart Circ. Physiol.* **282**(2), H466–H474 (2002)
8. V.L. Voeikov and V.I. Naletov, in: *Optical Diagnostics of Biological Fluids III,* edited by A. V. Priezzhev, T. Asakura, and J.D. Bries (SPIE Proc., **3252**, San Jose, CA, 1998), pp. 140-148.
9. K.K. Peachman, D.S. Lyles, and D.A. Bass, Mitochondria in eosinophils: Functional role in apoptosis but not respiration, *Proc Nat Acad Sci USA,* **98**(4), 1717–1722 (2001).
10. P. Wentworth Jr., L.H. Jones, A.D. Wentworth, X. Zhu, N.A. Larsen, I.A. Wilson, X. Xu, W.A. Goddard 3rd, K.D. Janda, A. Eschenmoser, and R.A. Lerner, Antibody catalysis of the oxidation of water *Science,* **293**(5536), 1806-1811 (2001).
11. V.L. Voeikov and C.N. Novikov, in: *Photon Propagation in Tissues III,* edited by D.A. Benaron and B. Chance (SPIE Proc, **3194**, San Remo, Italy, 1997) pp. 328-333.
12. V.L. Voeikov, R. Asfaramov, E.V. Bouravleva, C.N. Novikov, and N.D. Vilenskaya, Biophoton research in blood reveals its holistic properties, *Indian J. Exp. Biol.,* **43**(May), 473-482 (2003)
13. V.L. Voeikov and V.I. Naletov in: *Biophotons.* edited by J.-Ju. Chang, J. Fisch, and F.-.A. Popp (Kluwer Academic Publishers, Dortrecht, The Netherlands, 1998) pp. 93-108.
14. V.L. Voeikov, I.V. Baskakov, K. Kafkialias, and V.I. Naletov Initiation of degenerate-branched chain reaction of glycin deamination with ultraweak UV irradiation or hydrogen peroxide, *Russ. J. Bioorganic Chem.,* **22**(1), 35-42 (1996).
15. H.S. Yim, S.-O. Kang, Y.C. Hah, P.B. Chock, and M.B. Yim, Free radicals generated during the glycation reaction of amino acids by methylglyoxal. A model study of protein-cross-linked free radicals, *J. Biol. Chem.,* **270**(47), 28228-28233 (1995).
16. V.L. Voeikov, V.V. Koldunov, and D.S. Kononov, New oscillatory process in solutions of compounds containing carbonyl- and amino groups, *Kinetics and Catalysis,* **42**(5), 606-608 (2001).
17. V.L. Voeikov, V.V. Koldunov, and D.S. Kononov, Long-duration oscillations of chemi-luminescence during the amino-carbonyl reaction in aqueous solutions, *Russ. J. Phys. Chem.,* **75**(10), 1443-1448 (2001).
18. A.G. Domrachev, Yu.L. Rodigin, and D.A. Selivanovsky, Mechanochemial activation of water splitting in liquid phase, *Doclady Acad. Sci. USSR,* **329**(3), 186-188 (1993).
19. S. Ikeda, T. Takata, M. Komoda, M. Hara, J.N. Kondo, K. Domen, A. Tanaka, H. Hosono, and H. Kawazoe, Mechano-catalysis -- a novel method for overall water splitting, *Phys. Chem. Chem. Phys.,* **1**(18), 4485-4491 (1999).
20. D. Porterfield, R. Corkey, R. Sanger, K. Tornheim, P.J. Smith, and B.E. Corkey, Oxygen consumption oscillates in single clonal pancreatic beta-cells (HIT), *Diabetes,* **49**(9), 1511–1516 (2000).

Chapter 12

TWO-DIMENSIONAL IMAGING AND SPATIOTEMPORAL ANALYSIS OF BIOPHOTON
Technique and applications for biomedical imaging

Masaki Kobayashi[1]

1. INTRODUCTION

Faint lights glowing in the dark, such as the blinking of fireflies, are very attractive. When bioluminescence is observed in living organisms, it seems fantastic and it stirs our curiosity. However, if we were to have extremely sensitive eyes that could detect a single photon, we would find that all of the living organisms in the world are actually shining like fireflies. Luminescence from living organisms is not a phenomenon that only applies to specific species such as fireflies, but is a general property that is possessed by all organisms. These emissions are ultraweak in intensity, with a range far removed from that of fireflies, and nobody can see them with the naked eye, although the phenomenon can be observed using highly sensitive photon detectors. The development of the photomultiplier tube in 1950s has gradually allowed these photon emission phenomena to be revealed[1]. Terms such as ultraweak photon emission and spontaneous ultraweak, low-level, or dark bio-/chemi-luminescence, etc. are now in general use to describe these phenomena, in order to distinguish them from general bioluminescence. The concept of the biophoton[2] is also commonly used to represent these phenomena. In this chapter, technologies for the determination and analysis of biophotons and several studies (chiefly based on the imaging of biophotons) for biological and medical applications aimed at diagnostic use are described.

In the last half-century, several studies have been carried out to explore various aspects of biophotons that have been observed in different organisms, particularly for the elucidation of their mechanisms and for the development of practical applications. Although a great deal of knowledge has been accumulated regarding biophoton phenomena,

---

[1] Masaki Kobayashi, Department of Electronics, Tohoku Institute of Technology, 35-1, Yagiyama-Kasumicho, Taihaku-ku, Sendai 982-8577, Japan, E-mail: masaki@tohtech.ac.jp

the weak intensity of the emissions has restricted the practical application of biophoton technology to just a few limited fields. In order to make further progress in the development of biophoton applications, sophisticated techniques for analyzing the faint emissions from a restricted number of photons are required. We have continued studying biophoton emission from a variety of viewpoints in the fields of biophysics, biochemistry, and biomedical engineering, and we have contributed to advances in the interpretation of their mechanisms and the development of applications, including analysis methods and instrumentation[3]. Here, I will review the modern technology used for biophoton analysis and also feasibility studies for a wide range of subjects to determine if there is any useful pathophysiological information that they can convey.

We have developed the following instrumentation systems: (1) A highly-sensitive photon counting apparatus designed for the improvement of long-term stability[4]. (2) A biophoton imaging apparatus for spatiotemporal analysis based on a two-dimensional photon counting technique. (3) An imaging apparatus using a highly sensitive charge coupled device (CCD). (4) An apparatus for the spectral analysis of biophotons using a set of colored-glass filters with a wide range of wavelengths covering from the ultraviolet to the near infrared regions[5]. (5) An apparatus for determining photon statistics and photon correlation characteristics based on a method that traces single-photon pulses from a detector[4,6]. Here I will describe the systems focused on the imaging apparatus for biophoton analysis based on the two-dimensional photon counting technique and a highly sensitive charge coupled device (CCD) imaging technique.

In terms of feasibility studies for biomedical applications, experimental results obtained from the measurement of plants and mammals through imaging and spatiotemporal characterization to clarify the relationship between biophotons and pathophysiological responses are described and discussed as follows. (1) The responses of plants to exogenous stress. (2) Determination of the spatiotemporal propagation of pathological responses under induced oxidative stress observed on a mouse body. (3) Biophoton emission from the brain of a rat that is associated with observed neuronal activity. (4) Biophoton imaging of a human body.

2. TWO-DIMENSIONAL DETECTION AND ANALYSIS TECHNIQUE OF BIOPHOTONS

2.1 Photon Counting Imaging and Spatiotemporal Analysis[7,8,9]

The imaging system for biophoton emission developed in our group consists of a two-dimensional photon counting tube with a large active area, a highly efficient lens system installed in a sample chamber, and an electronic apparatus for identifying the two-dimensional spatial and temporal photoelectron information. A block diagram is shown in Figure 1. The photon counting tube (Model IPD 440, Photek, Ltd., UK), which was installed in the vacuum chamber, had a photocathode measuring 40 mm in diameter with spectral sensitivity (S-20) operating at a wavelength ranging from 350 to 900 nm, and with a quantum efficiency of 9% at 500 nm, 5.5% at 600 nm, and 1.3% at 800 nm.

The tube dark count was less than 76 counts/s over the whole effective area with cooling at $-35°C$. Spatial resolution of the tube, which was determined by the readout precision of the resistive anode incorporated into the photon counting tube, was

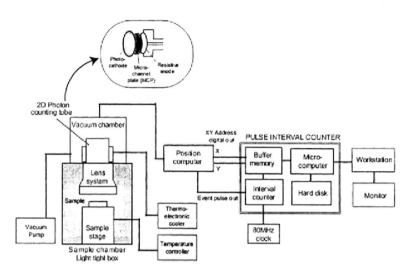

Figure 1. A block diagram of biophoton imaging system characterized with a highly sensitive imaging and spatiotemporal analysis for biophoton emission. Timing pulses in photoelectron event from a position computer operate a pulse counter to measure pulse-to-pulse interval based on an 80 MHz clock. Time interval data and position data are transferred to a buffer memory and finally stores in a hard disk continuously.

approximately 200 μm.

A specially designed lens system (Fujii Optical Co., Tokyo, Japan) had a 90 mm aperture with a 0.5 N.A. (numerical aperture) and 1.0 magnification corresponding to an image size of 25 by 25 mm or 1/3 with 75 by 75 mm image size. Output pulses from the resistive anode were fed to a position computer (IPD controller, Photek) to determine the X-Y position of each photoelectron event. These data (9 bits, 2 channels) were consecutively transferred to a pulse-interval counter (Tohoku Electronic Industrial Co., Sendai, Japan) and stored with timing data of the event, which is represented as the time interval between two successive photoelectron pulses. The time resolution of events is restricted by the position-computer to a pulse-pair resolution of 10 μs.

After data acquisition, the data were transferred to a workstation for reconstruction of the photon counting images and analysis of spatiotemporal properties, which demonstrated the intensity kinetics in regions of interest or space-time correlation of the photoelectrons.

Images were processed with data correction for the spatial distribution of dark counts and Gaussian smoothing. Spatial resolution on the processing was adapted with the sample intensity. Minimum detectable radiant flux density, which was defined with dark counts from the detector, quantum efficiency of the photocathode, photoelectron collection efficiency, and light collection efficiency of the lens system, was experimentally evaluated and found to be 9.90×10^{-17} W/cm^2 (= 314 photons/s/cm^2 at a wavelength of 670 nm) under the condition of spatial resolution of 2.3mm in the case of 1.0 magnification. A single photoelectron of the detector was estimated to correspond to 221 photons emitted onto the sample surface.

Raw data accumulated in the pulse interval counter are expressed by a set of sequences $\{(x_1, \xi_1), (x_2, \xi_2), ..., (x_N, \xi_N)\}$, where ξ_i is the arrival time-interval between the $(i-1)$-th and the i-th photoelectron and x_i is the two-dimensional location of the i-th photoelectron; and N is the total number of photoelectron pulses. Initially, a photon counting image, expressed as $n(x, y)$, which is the number of photoelectron pulses at position (x, y) during a total measurement period, is constructed. After setting regions of interest $(r_j = \{[x_{j1}, x_{j2}], [y_{j1}, y_{j2}]\})$ on the image plane, the number of photoelectron pulses per unit time T (observation time) in each region are calculated as $n(r_j, t, T)$ to extract the intensity time course. The spatial and temporal correlations of emission intensity $\langle I(r_1, t) I(r_2, t+\tau) \rangle$ are calculated from the photoelectron correlation that is derived from the photoelectron detection probability at a space-time point (r_1, t) and detection probability after time τ at r_2, represented by the conditional probability $P_c(r_2, \tau)$ of two successive photoelectron pulses. The relationship is expressed as

$$\langle I(r_1, t) I(r_2, t+\tau) \rangle = \alpha \langle n(r_1, t, T) \rangle P_c(r_2, \tau)$$

where the brackets $\langle \rangle$ denote ensemble averaging and α, a constant.

The advantage of this system is the arbitrary selection of spatial and temporal dimensions (r_j and T, τ), which can be achieved from a single measurement data point instead of making a fresh measurement for each change in the spatial or temporal dimensions.

2.2 CCD Imaging[10]

A highly sensitive CCD camera system with a thinned back-illuminated CCD detector, cooled to below $-100°C$, is also capable of biophoton imaging. This type of CCD camera, which has been developed for astronomical observation, is equipped with a cooling system using liquid nitrogen or a closed-cycle-mechanical cryogenic cooler. A multi-phase pinned (MPP) architecture is also included for the reduction of the dark current. The operation is performed with an inverted bias voltage on the device during integration and line readout, resulting in a reduction in the dark current by a factor of 100-1000. However the full potential-well capacity is reduced when using the MPP mode, causing a lessening of the intensity dynamic range. The readout rates are restricted in order to reduce the readout noise, occurring in the buffer amplifier built into the CCD chip. Thus this type of camera is referred to as a slow-scan CCD camera. Although the slow-scan limits the time resolution of a measurement to the order of tens of minutes, this is not insignificant for biophoton imaging, considering the integration time.

When considering the minimum detectable optical power in comparison with a two-dimensional photon counting camera, it is necessary to consider the difference in the definition of signal-to-noise ratio. The quantum efficiency of the photocathode in a 2D photon counting tube is low, being approximately 20% at peak wavelength of 400 nm and less than 5% in the red or near-IR regions. By contrast, in the case of a back-illuminated type CCD, the quantum efficiency at the peak wavelength is approximately 70%, even under the cold conditions and the sensitivity ranges between 400 nm to 900 nm.

The minimum detectable optical power at each pixel of a 2D photon counting tube (2D PMT) with a signal to noise ratio of unity can be defined as

$$(P_{min})_{2DPMT} = \frac{\sqrt{(N_{d1}T)}\,hc}{\lambda\,\eta_1 T S}$$

where N_{d1} is the number of dark counts per unit time in a single element of 2D PMT, T is the measurement time, h is Planck's constant, c is the velocity of light, η_1 is the quantum efficiency of the PMT photocathode, λ is the wavelength, and S is the area of each pixel. In the case of a CCD camera, the minimum detectable power is defined as

$$(P_{min})_{CCD} = \frac{\sqrt{(N_{d2}T + N_r^2)}\,hc}{\lambda\,\eta_2 T S}$$

where N_{d2} is the number of electrons contributing to the dark current per unit time at a single element of the CCD, η_2 is the quantum efficiency of the CCD, and N_r represents the electron root mean square (RMS) of readout noise of amplifier circuit per pixel. From the above equations it can be deduced that, if the CCD is operated at extremely low dark current and longer exposure times to reduce the contribution of the readout noise, the signal to noise ratio of the CCD camera would be superior to the conventional 2D PMT under similar conditions of wavelength and measurement time. Another noise to be considered results from spurious events induced by secondary particles of cosmic rays in the atmosphere, such as muons. Typically the rates of these events are known approximately 2 counts/cm^2/min with depending on the solar activity. When considering the characteristics of these events, resulting in extremely high count levels, localized in a small number of pixels, they can be easily reduced through filtering by image processing[11].

Considering the experimental equipment arrangements, the imaging format of CCD (TK1024AB2-G1; SITe, OR, USA) incorporated in CCD camera system (ATC200C; Photometrics, Arizona, USA) comprised 1024 x 1024 pixels for a full frame, with the size of each pixel being 24 x 24 µm. Dark current of the device was 0.225 e$^-$/hr/pixel (= 6.25x10^{-5} e$^-$/s/pixel) at $-120°$C in MPP mode and the readout noise was 3.3 e$^-$ RMS/pixel. Dependent on the intensity of the emission, the experimental spatial resolution of the CCD was regulated using a binning mode. Quantum efficiency of the system was 73% at 700 nm. When comparing against the two-dimensional photon counting system, the selection is required of an imaging device, covering a suitable emission spectrum and integration time, dependent on the emission intensity required for optimum imaging of biophoton. Recently, there have been significant advancements in highly-sensitive cooled CCD camera systems, offering a CCD system which includes a mechanical cryogenic cooler capable of $-110°$C, equipped with the high resolution back-illuminated type of CCD.

3. CHARACTERIZATION OF BIOPHOTON PHENOMENA FOR BIOLOGICAL MEASUREMENTS AND APPLICATIONS

Biophoton emission originates in the electronically-excited states of the constituents of living cells, which are generally associated with the presence of an oxidative metabolism that accompanies the production of reactive oxygen species (ROS). During the normal energy metabolism, cellular respiration (a reaction in the electron transfer chain of the inner mitochondrial membrane) participates in ROS production, which is especially facilitated under the highly-reduced state of an electron transfer chain.

Biophoton emission reflects the pathophysiological state with respect to energy (ATP) production and the susceptibility to oxidative stress, which is derived from the excessive production of ROS or a lack of activity for antioxidant protection. Ultraweak photon emission at the subcellular level, such as that from isolated mitochondria[12, 13, 14] and at the cellular level, such as that from cultured carcinoma cells[15, 16, 17] suggest a relationship between photon emission intensity and metabolic activity. Boveris et al. characterized photon emissions from a variety of mammalian organs in an in vivo investigation of the reactions of radicals through lipid peroxidation. Many pioneering studies have suggested the potential usefulness for noninvasive monitoring of oxidative metabolism and oxidative damage to living tissue[18, 19] under physiological and pathological conditions.

In this section, I will describe feasibility studies for biological measurements to extract physiological or pathological information for various subject organisms based on two-dimensional imaging.

3.1 Plants[7]

A biophoton image of an intact sample, consisting of etiolated soybean seedlings (*Glycine max*) that have been dark adapted for 3-4 days, is displayed in Figure 2 (a) with a sample photograph under weak light illumination in Figure 2 (b). The length of the root system was approximately 30-40 mm during the growth period. Although this image was taken with 30 minutes integration time, using the CCD imaging system, the spatial distribution and patterns of biophoton emission intensity in the hypocotyl area are clearly observed.

The hypocotyl area, indicated by strong emission, is considered to be responsible for cell growth and respiration, hence increased emission intensity suggests that biophotons carry information regarding cellular metabolic activities.

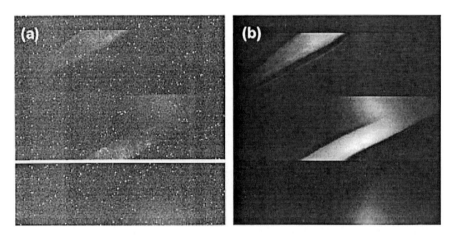

Figure 2. A biophoton image of an intact soybean seedling observed with using a highly sensitive CCD imaging system (a) Biophoton image obtained with integration time of 30 minutes. (b) Image of the sample taken under weak light illumination.

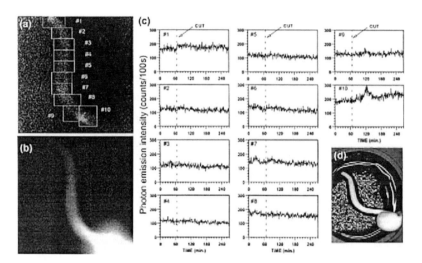

Figure 3. Spatiotemporal variation of biophoton observed after stimulation of root-tip excision of soybean under germination. (a) Biophoton image obtained with integration time of 4.5 hours. (b) Image of the sample taken under weak light illumination. (c) Time courses of biophoton emission intensity in different regions indicated as white squares on the Figure 3 (a). (d) The photograph of the sample after the measurement.

The highly resolved image detected by the CCD imaging system has the potential of mapping the activity of cellular energy metabolism. The determination of the physiological responses of plants has been studied by means of the analysis of the spatiotemporal properties of biophoton emission.

The biophoton responses from a soybean root were analyzed under various external stimulation conditions. There follows a typical result of observing the response after stimulation by root tip excision (Figure 3). In Figure 3 (c), the variation of the biophoton emission on excision of the root tip is displayed over time of the intensity in the selected regions defined in Figure 3 (a), indicating the integrated image over the total measurement time of 4.5 hours. The emission intensity of the cut region (Figure3 c #1) increased and remained at a high level for 3 hours. Significant changes in photon emission intensity were observed at a remote position (Figure3 c #10). The manifestation of a response by the soybean seedling to injury, as observed by an increase in photon emission at a position remote from the injury site is very interesting. This region, referred to as the hypocotyls, is known to be highly active in cellular respiration; therefore it is suggested that the temporary enhancement of photon emission could be a reflection of the increased metabolic activity, associated with the cellular respiration induced by the external stimulation. It is suggested that the biophoton emission from a mechanically-injured soybean root involves a contribution from the endogenous H_2O_2-peroxidase system[20, 21], and it is postulated that the phenomenon reflects a defensive response within the plant to seal off the wound and to generate new tissue for wound-healing through the activation of the peroxidase system.

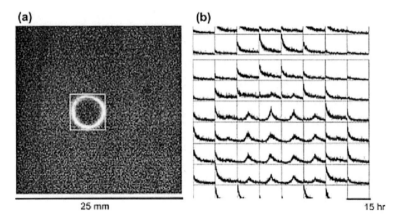

Figure 4. Spatiotemporal variation of biophoton observed after heat stimulation on a cowpea leaf by a CO_2 laser beam. (a) Biophoton image obtained with integration time of 30 minutes after the treatment. (b) Time courses of biophoton emission intensity over 15 hours after the heat stimulation in different regions indicated as white squares on the Figure 4 (a).

This demonstrates the potential usefulness of the analysis of biophotons dynamics to monitor the response of a living system to environmental stimuli. Reactions occurring at a remote site may indicate the holistic response of the plant. Living systems maintain a complex order, both at the cellular level and at the macroscopic level. It is natural to assume that any form of external stimulus, be it physical, chemical or pathological, would disturb this order and manifest itself at different locations.

Figure 4 demonstrates biophoton response after heat stimulation. The experiment was carried out using extracted cowpea leaves, following an application of a CO_2 laser beam on surface. The surface temperature of a laser spot was monitored with thermography during heat stimulation. As shown in Figure 4 (b), indicating the spatiotemporal characteristics of biophoton emission over 15 h after the treatment, a distinctive change of spatial distribution of photon emission intensity with time can be identified. The surface temperature at the center was estimated to be 70°C with a laser power of 400 mW applied for 10 seconds. Primarily, the peripheral region of the spot exhibited the strong emission, followed by decay. On the other hand, in the center region, the photon emission intensity increased temporally and peaked at approximately 10 hours after application, implying a difference in the photon emission mechanism occurring in the center and the periphery. The monotonic decrease observed in the peripheral region is speculated to be a reflection of physiological response against heat stress to protect and repair the damaged tissue, and the augmentation observed in the center region may indicate induction of chemical reactions including radical reactions, accompanied by necrosis which was characterized by browning.

Observations introduced in here are suggestive of the effectiveness of the spatiotemporal analysis, depicting the dynamics of biological response, together with the potential usefulness of the analysis of biophoton dynamics as a response of a living system to environmental stimuli. The results of this present study show that the method could add a new dimension to the non-invasive study of the response of plants to injury or diseases, and could also contribute to the study of mechanisms of signal transfer within a living system.

3.2 Mammal[22]

We have attempted to establish a technique whereby pathophysiological information in mammals can be visualized in vivo based on biophoton imaging. Determination of the spatiotemporal distribution of biophotons using a two-dimensional photon counting technique could also provide the kinetics of the pathological and/or physiological states. Here, I will introduce some experiments that were carried out using mice to visualize the spatiotemporal propagation of oxidative stress and oxidative injury occurring in internal organs.

Paraquat is known to generate superoxide through the radical reaction of paraquat ions with oxygen, and its ingestion induces oxidative injury of the internal organs. We examined the changes in the biophoton images observed on the body surface after the administration of paraquat. A nude mouse was placed in the sample chamber under anesthesia. After the oral administration of paraquat, its biophoton image was observed in the supine position for 15 hours. Figure 5 shows a typical result, indicating the temporal changes of the biophoton images. Figures 5 (a)–(c) represent the images observed at 2, 5, 7 hours after administration respectively, with 1 hour integration. The small graphs in Figure 5 (d) represent the spatiotemporal characteristics of the biophoton emission expressed with the temporal changes of the intensity with a time resolution of 15 minutes for each region, which were divided into the positions indicated on the profile of the mouse body. The results indicate a distinctive augmentation of the intensity centered on the area corresponding to the stomach 2-3 hours after administration and at the intestine after 5-10 hours. Control experiments using untreated mice under normal conditions showed no remarkable changes in the intensity distribution over the same period. This augmentation of the intensity implies the propagation of the affected part by oxidative injury. We also examined the site of origin of the photon emission to determine whether or not it originated inside the body, i.e., in the internal organs. We found that the majority of the detected photons were derived from the internal organs and/or the hypodermis, with a lesser contribution from the skin surface.

Microscopic observations of cellular damage to tissue that was removed from the sacrificed mouse after the measurement showed damage to the mucous membranes in the stomach and in the intestine. Although analysis in detail is necessary for pathological diagnosis, the preliminary results derived from these experiments suggest the potential for the application of spatiotemporal analysis of biophotons in pharmacokinetic and pathological investigations. For example, it may applicable to the assessment of oxidative stress induced by adverse reaction to a drug.

Biophoton imaging was also studied in cancer-transplanted mice. Significant enhancement of the emission intensity was observed according to the growth of the tumor, indicating that it correlated with the growth rate and viability of the cancer cells[17, 23]. This observation offers a valuable application for the evaluation of malignancy and for pharmacological studies on the effects of anti-cancer drugs.

3.3 Rat brain[9]

In this section, two-dimensional imaging of the biophoton emission from a rat's brain, detected in vivo over the skull, is demonstrated. The physiological properties of the emission associated with metabolic activity, through simultaneous measurement of electroencephalographic (EEG) activity, are described. Analyses of the mechanisms of photon emission are also presented by using spectral analysis of in vitro brain slices.

3.3.1 Correlation Between Photon Emission Intensity and EEG Activity

Biophoton imaging of rat's brain was performed under anesthesia and artificial ventilation. After the incision of skin to exposure the skull, measurements have been carried out, under the atmospheric condition of nitrogen to eliminate the artificial chemiluminescence by autooxidation of the exposed tissue surface, with various physiological conditions. An example of the time course of biophoton emission intensity and simultaneously measured EEG activity represented by the theta wave component of the EEG power spectrum is displayed in Figure 6 (a). Temporal changes of the photon emission intensity were relatively comparable to the theta wave activity. Changes in the spatial pattern of emission were also observed from images in the respective time regions, as shown in Figure 6 (b). Figures 7 (a) and (b) show the results of correlation analysis between photon emission intensity and theta wave activity, represented by 30-minute integration under a different condition of skull treatment. Figure 7 (a) shows the result obtained from four animals where the parietal bones were removed and (b), that obtained over the skull from eleven animals. Both figures have been composed by superimposing independent measurements. Although emission intensities observed through the skull are approximately one-half of those with the parietal bones removed, both results support the correlation between photon emission intensity and the theta wave component of the EEG power spectra, with statistical significance (p<0.001). This result implies the relationship between biophoton emission intensity and metabolic activity of neural cells as interpreted with expression of energy metabolism.

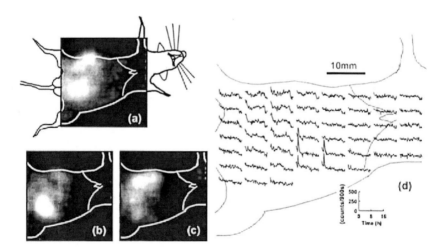

Figure 5. Spatiotemporal analysis of biophoton emission of a mouse under oxidative stress induced by paraquat administration. (a) An image obtained 2 hours after the administration. (b) 5 hours after the administration. (c) 7 hours after the administration. (d) Spatiotemporal properties of biophoton intensity. Each graph shows time course of emission intensity at the indicated position. All images of biophoton were taken with 1-hour integration.

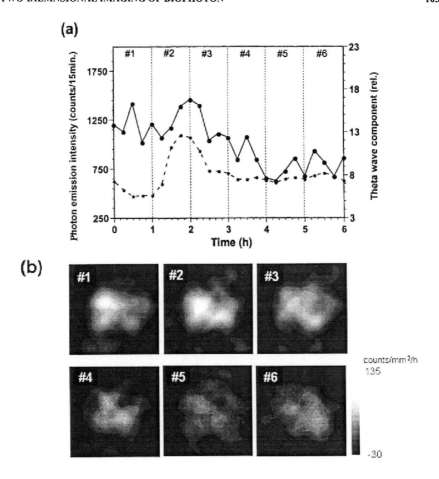

Figure 6. (a) An example of temporal changes in biophoton emission intensity from a rat's brain and theta wave component of EEG power spectral density (ratio of component expressed in %, rel.). (b) Sequential images of photon emission for the time regions of #1-#6, indicated in (a).

3.3.2 Ultraweak Photon Emission Spectra of Brain Slices

In order to elucidate the mechanism for biophoton emission from a rat's brain, we have performed spectral analysis of extracted brain slices. The slices were prepared after decapitation under diethyl ether anesthesia. One hemisphere of the brain was frontally-sectioned into slices of 500-μm thickness using a slicer. We obtained 10 sheets of slices, which were placed in a quartz chamber containing circulating artificial cerebrospinal fluid (ACSF) that was bubbled with a mixture of 95% oxygen and 5% CO_2. A reference spectrum observed under normal conditions is shown in figure 8 (a). This shows an emission spectrum ranging from 500 to 800 nm, with peaks around 530 nm and 610 nm and

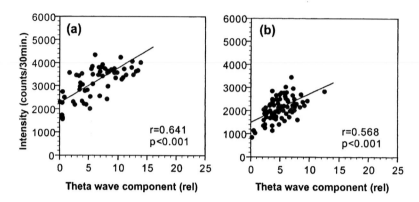

Figure 7. Correlation analysis between biophoton emission intensity and theta wave component of EEG power spectra represented by 30-minute integration. (a) Correlation after having the bilateral bones removed and (b) without removing the bone. Correlation coefficients and statistical significance are indicated in the figures.

a shoulder at 670 nm. The dashed line in Figure 8 (a) shows the spectral pattern measured under glucose deprivation in the ACSF. As can be seen in Figure 8 (b), which shows temporal changes in the intensity before and after glucose deprivation, the photon emission intensity in the case of glucose deprivation was depressed by approximately 20% in comparison with normal (indicated by glucose (+)) conditions[9]. However, no significant differences in shape could be seen between the two spectral patterns, except for a minor depression of the intensity in the wavelength region between 600 nm and 700 nm. We also carried out analyses to compare the reference spectrum with spectra taken under conditions where glutamate or rotenone had been added for neuronal activation or for inhibition of the mitochondrial electron transfer chain, respectively. A temporary increase in the emission intensity was observed in both cases after the treatments, suggesting a relationship between the photon emission and activation of the energy metabolism through the electron leakage in the respiratory chain[9]. Once again no remarkable changes were recognized in the spectral patterns in either case.

In order to investigate the emission mechanisms, we compared the spectral patterns of the slices with the chemiluminescence from unsaturated fatty acids, which were observed under gradual autooxidation with pure oxygen at 37°C. The three lines in Figure 9 show the spectra of oxidized linoleic acid, linolenic acid and arachidonic acid, respectively. Linoleic acid exhibited temporal changes in its spectral pattern during autooxidation. The early stages of the autooxidation featured 3 peaks at around 530 nm, 630 nm and 700 nm, with spreading over 450-750 nm (data not shown). In the latter stages, enhancement of the emission intensity was observed, centered on the peak at 530 nm. In the case of linolenic acid, the spectral pattern ranged from 600-800 nm, with a dominant peak at 650 nm. Although a temporal increase in the photon-emission intensity during the autooxidation process was observed, no notable changes occurred in its spectral pattern. In the case of arachidonic acid, the chemiluminescence during the autooxidation showed a constant intensity in terms of photon emission. Its spectral pattern was also unchanged during the measurement period and the emission spectrum, which ranged between 450-700 nm

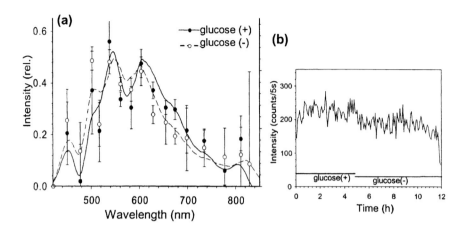

Figure 8. (a) A biophoton emission spectrum of brain slices (solid line) in comparison with the conditions of glucose deprivation (dash line). (b) Temporal changes of the emission intensity of slices under the condition with glucose and without glucose in ACSF.

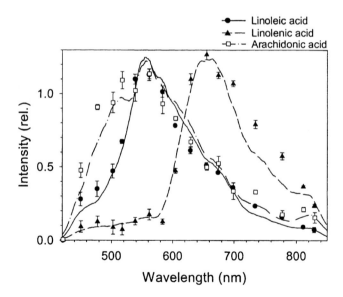

Figure 9. Comparison of chemiluminescence spectra among linoleic acid (solid line with filled circle; later stage of oxidation), linolenic acid (dash line with filled triangle) and arachidonic acid (dash-dot line with open square) during autooxidation promoted with pure oxygen under 37°C.

included peaks at 490 nm and 530 nm. The spectral pattern was similar to that of the latter stages of oxidation of linoleic acid, except in the emission region at around 450-520 nm.

A comparison of these results with those of the brain slices suggests that arachidonic acid and linoleic acid or their derivatives may be candidates for the excited species involved in the biophoton emission detected from rat brains. The pathway for excitation may include radical reactions through lipid peroxidation with ROS production. However, the peak obtained from the brain slices at 610 nm does not correspond to any of the oxidized fatty acids, indicating the existence of other sources of the emission. We can speculate that these might be oxidative derivatives from proteins or other molecules.

3.4 Human body[22]

In order to survey the application of biophotons to the measurement of human subjects, we tried to evaluate potential methods for charactcrizing thc biophoton image of a human body. We also investigated ultraweak photon emission of samples originated in human subjects; blood plasma, urine, and breath were examined[5]. The biophoton emissions detected from a palm, a face and from the upper part of the waist were examined using a two-dimensional photon counting imaging system and a highly sensitive CCD imaging system. Biophoton images that were determined after 30 minutes observation are displayed in Figure 10 with the image of the subjects under weak light illumination. These images were taken after dark-adaptation for 10 minutes to minimize the effect of delayed fluorescence. The images display the characteristic distribution of the intensity of biophoton emission on the surface of the body. As shown by the image of the upper part of the waist, the photon emission intensity around the face is higher than it is around the body. This may represent a difference in the condition of the skin due to exposure to environmental light. In the case of the face, the intensity in the central area around the nose

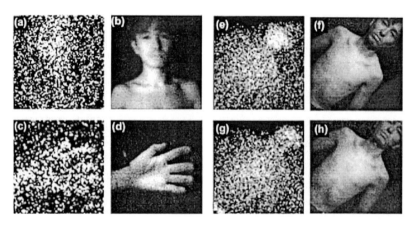

Figure 10. Biophoton images of a human subject. (a) Biophoton image of a face taken as shown in the photograph (b). (c) Biophoton image of a palm taken as shown in the photograph (d). (e), (g) Biophoton images of the upper part of the waist taken as shown in photographs of (f) and (h), respectively. (a) and (c) were measured with using the two-dimensional photon counting imaging apparatus, and (e) and (g) were measured with the CCD imaging system. All of biophoton images were taken with observation time of 30 minutes

is observed to be higher. This could be caused by secretions on the skin surface or more speculatively the influence of porphyrin compounds secreted by indigenous bacteria on the skin. In order to estimate the contribution to the photon emission intensity of the autooxidation of secreted sweat or sebum on skin surface, the dependence of the photon-emission intensity on atmospheric oxygen was investigated.

We compared the photon emission intensity detected from the palm of a human hand under aerobic and anaerobic conditions by filling the measurement chamber with either pure oxygen or pure nitrogen. The result, which shows the remarkable dependence of the photon-emission intensity on the concentration of oxygen, suggests the contribution of lipid peroxidation occurring on the skin surface, probably including the peroxidation of secreted fluids. The photon-emission intensity from a palm under anoxia was depressed by approximately 25% compared with the value in air and by 50% under pure oxygen. This implies that the photons detected on the body surface contain an atmospheric-oxygen-independent component, originating from sub-epidermal or dermal cells, or perhaps from other cells below the basal layer. The determination of this component may lead to information for evaluating the susceptibility to oxidative stress of dermal cells, which may in turn lead to diagnostic applications for skin medicine. Biophoton imaging is now under investigation to explore applications under various environmental conditions, including exposure to ultraviolet (UV) irradiation, chemical agents and other physical stimuli. Many other studies aimed at the medical application of biophotons from the skin surface have been reported from a variety of viewpoints, including on the concept of delayed luminescence after UV light exposure[24, 25].

4. CONCLUSION

In this chapter, our studies focused on techniques and instrumentation for highly sensitive and precise determination of biophotons through imaging and their applications, investigated from various viewpoints, are discussed. Applications based on these techniques are introduced over a range of subjects, from the level of microbes to that of humans. Although the detailed mechanisms of biophoton emission in each case have not yet been clarified sufficiently, the experimental results introduced here show a high potential for analyzing biophotons for the quantification of physiological and pathological information. The most valuable merit of analyzing biophotons is that it is apparently a non-invasive technique, and it cannot be influenced in the same way as other methods that use extrinsic physical probes. However, the weakness of the intensity of biophoton, which may restrict its application, requires a breakthrough in analytical technique. In that sense, I believe that our technologies for the physical analysis of biophotons could form the basis for a novel stage in biophoton research. In particular, the statistical characterization of biophotons with spatiotemporal determination will provide a novel approach by which biological order emerged with the interaction of biological reactions can be characterized, even when the photon density is very low.

5. ACKNOWLEDGEMENTS

These works were carried out in the Inaba Biophoton Project, ERATO program, Japan

Science and Technology Corporation (1986-1991), and the Biophotonics Information Laboratory Co., Ltd. (1993-1999).

6. REFERENCES

1. L. Colli and U. Facchini, Light emission by germinating plants, *IL Nuovo Cimento* **12**, 150-153 (1954).
2. F. A. Pop (Ed.), Special issue on biophoton, *Experientia* **44**, 543-600 (1988).
3. H. Inaba, New bio-information from ultraweak photon emission in life and biological activities: biophoton, in: *Modern Radio Science 1990*, edited by J.B. Andersen (Oxford Univ. Press, Oxford, 1990), pp.164-184 and references cited therein.
4. M. Kobayashi and H. Inaba, Photon statistics and correlation analysis of ultraweak light originating from living organisms for extraction of biological information, *Appl. Otp.* **39**(1), 183-192 (2000).
5. M. Kobayashi, M. Usa, and H. Inaba, Highly sensitive detection and spectral analysis of ultraweak photon emission from living samples of human origin for the determination of biomedical information, *Trans. of SICE* **E-1**, 214-220 (2001).
6. M. Kobayashi, B. Devaraj, and H. Inaba, Observation of super-Poisson statistics of bacterial (*Photobacterium phosphoreum*) bioluminescence during the early stage of cell proliferation, *Phys. Rev. E* **57**(2), 2129-2133 (1998).
7. M. Kobayashi, B. Devaraj, M. Usa, Y. Tanno, M. Takeda, and H. Inaba, Development and applications of new technology for two-dimensional space-time characterization and correlation analysis of ultraweak biophoton information, *Frontiers Med. Biol. Engng.* **7**(4), 99-309 (1996).
8. M. Kobayashi, T. Takeda, K-I. Ito, H. Kato, and H. Inaba, Two-dimensional photon counting imaging and spatiotemporal characterization of ultraweak photon emission from a rat's brain in vivo, *J. Neurosci. Method* **93**(2), 163-168 (1999).
9. M. Kobayashi, M. Takeda, T. Sato, Y. Yamazaki, K. Kaneko, K-I. Ito, H. Kato, and H. Inaba, In vivo imaging of spontaneous ultraweak photon emission from a rat's brain correlated with cerebral energy metabolism and oxidative stress, *Neurosci. Res.* **34**(2), 103-113 (1999).
10. M. Kobayashi, B. Devaraj, M. Usa, Y. Tanno, M. Takeda, and H. Inaba, Two-dimensional imaging of ultraweak photon emission from germinating soybean seedlings with a highly sensitive CCD camera. *Photochem. Photobiol.* **65**(3), 535-537 (1997).
11. C.D. Mackey, The role of charge coupled devices in low light level imaging, in: *A Technical Guide from Astromed Limited*, Issue 1, pp.1-25 (1991)
12. T.B.Suslova, V. I. Olenev, and Y.A. Vladimirov, Chemiluminescence coupled with the formation of lipid peroxides in biological membranes-I. Luminescence of mitochondria on addition of Fe^{++}, *Biofizika* **14**, 540-548 (1969).
13. E.Cadenas, A.Boveris, and Chance, B, Low-level chemiluminescence of bovine heart submitochondrial particles, *Biochem. J.* **186**, 659-667 (1980).
14. E. Hideg, M. Kobayashi, and H. Inaba, Spontaneous ultraweak light emission from respiring spinach leaf mitochondria, *Biochim. Biophys. Acta.* **1098**, 27-31 (1991).
15. T.G. Mamedov, G.A. Podov, and V.V. Konev, Ultraweak luminescence of various organisms, *Biofizika* **14**, 1047-1051 (1969).
16. M. Takeda, Y. Tanno, M. Kobayashi, M. Usa, and H. Inaba, A novel method of assessing carcinoma cell proliferation by biophoton emission, *Cancer Lett.* **127**(1-2), 155-160 (1998).
17. T. Amano, M. Kobayashi, B. Devaraj, M. Usa, and H. Inaba, Ultraweak Biophoton Emission Imaging of Transplanted Bladder Cancer. *Urol. Res.* **23**, 315-318 (1995).
18. E. Cadenas, A. Boveris, and B. Chance, Low-level chemiluminescence of biological systems, in: *Free radicals in biology, Vol. VI*, edited by W. A. Proyor (Academic Press, New York, 1984), pp. 211-242.
19. A. Boveris, E. Cadenas, R. Reiter, M. Filipkowski, Y. Nakase, and B. Chance, Organ chemiluminescence: Noninvasive assay for oxidative radical reactions, *Proc. Natl. Acad. Sci. USA* **77**, 347-351 (1980).
20. M.L. Salin, K.L. Quince, and D.J. Hunter, Chemiluminescence from mechanically injured soybean root tissue. *Photobiochem. Photobiophys.* **9**, 271-279 (1985).
21. S. Suzuki, M. Usa, T. Nagoshi, M. Kobayashi, N. Watanabe, H. Watanabe, and H. Inaba, Two-dimensional imaging and counting of ultraweak emission patterns from injured plant seedlings, *J. Photochem. Photobiol. B:Biol.* **9**, 211-217 (1991).
22. M. Kobayashi, Highly sensitive detection of reactive oxygen species originating in living body based on

biophoton detection, Optronics, No.12, pp. 139-147 (1999) (in Japanese).
23. M. Takeda, M. Kobayashi, M. Takayama, S. Suzuki, T. Ishida, K. Ohnuki, T.Moriya, and N. Ohuchi, (2004) (submitted).
24. G. Sauermann, W. Mei, U. Hoppe, and F. Stab, Ultraweak photon emission of human skin in vivo: Influence of topically applied antioxidants on human skin, in: *Methods in Enzym.* **300**, edited by L. Packer (Academic Press, New York, 1999), pp. 419-428.
25. S. Cohen and F. A. Pop, Biophoton emission of the human body, *J. Photochem. Photobiol. B:Biol.* **40**, 187-189 (1997).

Chapter 13

ULTRAWEAK PHOTON EMISSION FROM HUMAN BODY

Roeland Van Wijk[1] and Eduard Van Wijk[1]

1. INTRODUCTION

In the 60's ultraweak photon emission from specific isolated tissues and organs was reported. The studies included contracting heart isolated from the thorax,[1] electrically stimulated nerves,[2] and muscles.[3,4] Presently, it is generally accepted that all living organisms emit ultra-weak light that can be recorded at tissue, cellular, and sub-cellular levels.[5,6]

Today many chemiluminescent systems are known, including reactions in the liquid, gas, and solid phases. Extensive descriptions in literature of these reactions started in the 70's of last century. Efficient excitation requires a large instantaneous energy release in a single excitation step. The radiation produced may be in the ultraviolet, infrared, or visible region of the spectrum. Spectral analysis has been extensively applied to identify the source of the photon emission in organs,[7,8] cell fractions,[9,10] and enzymatic reactions and biochemical processes involving free radicals and lipid peroxidations.[8,11]

Studies on photon emission from human body started at the end of the 1980's.[12-15] For technical reasons, most of these studies targeted the human hand. The emission from the hand was in the range of a few photons per second per cm^2. The majority of measurements reflected a low signal-to-noise ratio. This did not allow accurate comparative analysis of different body locations.

A major development in this field was the construction of a special device that allowed manipulation in three directions over the human body of a highly cooled and very sensitive photo-multiplier tube with high signal-to-noise ratio. With this device Cohen and Popp[15] examined both the palms of the hand and the forehead of an individual person, daily, over several months. They demonstrated long-term biological rhythms of spontaneous emission. Particularly interesting was the observation that frequently the bilateral emission from hands was higher in summer than in other parts of the year.

---

[1] Roeland Van Wijk, International Institute of Biophysics, Ehemalige Raketenstation, Kapellener Strasse, 41472 Neuss, Germany.

The construction and high sensitivity of this movable photomultiplier should allow systematic scanning of the photon density over different parts of the human body. This application can lead to reliable data only when a few major questions have been answered. What time is required after entering the darkroom before a reliable level of spontaneous emission can be recorded? How stable is spontaneous photon emission at a single body location. What is the shortest recording time for a location that will preserve valid data averaging? Does the pattern change during the day? Does photon emission over different skin locations reflect a symmetrical pattern? What is the wavelength spectrum of this photon emission? What biochemical reactions can be associated with the photon emission.

This chapter will discuss the applicability of the moveable device for multi-site registration of spontaneous photon emission and a protocol for analyzing the wavelength spectrum of the ultra-weak photon emission of the hand for the ultraviolet and visible light range (200-650 nm).

2. SPONTANEOUS PHOTON EMISSION AND DELAYED LUMINESCENCE FROM HUMAN SKIN

A special dark room is utilized with a count rate of less than 0.1 photons s^{-1} cm^{-2}. The inner size of this room has the following dimensions: 2 m x 1.5 m x 2 m. The photo-multiplier (EMI 9235 QB, selected type) in this room is designed for manipulation in three directions.[15] It is a 52 mm diameter, specially selected low noise end window photomultiplier. The spectral sensitivity range is 200-650 nm. It is mounted in a sealed housing under vacuum with a quartz window, and is thermoelectrically maintained at a low temperature of $-25°C$ in order to reduce the dark current (noise). An additional ring at the front port of the photo-multiplier tube allows the recording of round areas of skin of 9 cm diameter.

The subject is positioned in the dark room and the photo-multiplier tube is placed directly above the part of the body to be measured. Since photon emission of skin can be due to previous exposure to light, care must be taken to avoid such delayed emission, which would interfere with estimation of spontaneous photon emission of the skin. It requires that the subject be shielded from ambient light for at least one hour before measurement.

Delayed luminescence is registered over the palm of the hand after light exposure. It is most extensively observed after exposure to bright sunlight, or blue and UV by an artificial light source. The requirement for shielding has been studied by mimicking the exposure of subjects to external light using an Osram Ultra Vitalux 300 W lamp as source of white and UV light. Delayed luminescence is induced by illumination of the skin for two min at a distance of 20 cm. The average time between end of excitation and start of measurement is approximately 7 sec. This time is needed to bring the hand into the measurement position in the dark room and then open the shutter. The lamp is placed outside the dark room since it influences the background values when placed inside the dark room.

The kinetics of fading of delayed luminescence is measured continuously for 4 h after the end of the illumination. Figure 1 illustrates a plot of intensity vs. time in a double logarithmic manner. The data highlights an almost straight line. This implies that the decay of delayed luminescence has a hyperbolic function. Therefore, the half-life of

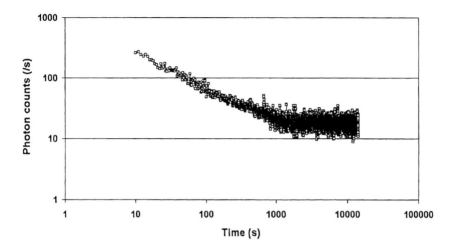

Figure 1. Delayed luminescence of human hand following irradiation with artificial sunlight.

the reduction in photon emission density is not constant; rather, it increases logarithmically over time.

One can conclude that delayed luminescence is able to interfere with the estimation of spontaneous emission for a long time after a subject enters the dark room.

In the example illustrated in Figure 1 the emission reaches a constant value after approximately 1 h. This emission is considered "spontaneous" emission over the palm of the hand. This value is neither influenced by the intensity nor by the wavelength of previous illumination. However, these two conditions determine the time required for dark-adaptation of the skin before spontaneous emission is measured in a reliable manner. Based on these studies it is advisable that subjects not be in bright sunlight for two hours and subsequently dark-adapted for 1 h prior to the measurement of photon emission from their skin.

3. DETERMINATION OF TIME SLOTS FOR RECORDING PHOTON EMISSION

For comparative studies of spontaneous emission at different locations, a duration of recording must be selected that allows, on the one hand, registration of as many skin locations as possible within a reasonable time, and, on the other hand, a high accuracy to reliably distinguish between intensity of different locations. However, these two variables act in an opposite manner. Therefore, to estimate an appropriate recording slot of time, a representative time series of photon counts of the palm of the hand from a healthy person is recorded and analysed.

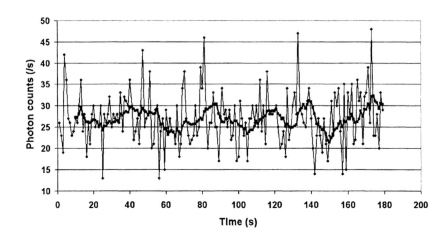

Figure 2. Spontaneous emission of human skin obtained during 3 min with a dwell time of 1 sec. Thin line: plot of original data. Thick line: same data after moving average.

Figure 2 illustrates a large variation in photon counts, and even irregular fluctuations in the order of tens of seconds. Such fluctuations impact the estimation of reliable average values and the number of time slots to be utilized for recording. From statistical analysis it is concluded that recording for 100 sec, with emission counts grouped for each sec, results in average values with low standard deviations, which do not significantly decrease when longer time series are used. Based on these studies the duration of each recording is commonly 2 min and consists of 120 time intervals of 1 s, unless otherwise stated.

4. TOPOGRAPHICAL VARIATION OF SPONTANEOUS PHOTON EMISSION

Figure 3 illustrates locations on the skin selected for recording of photon emission. The 29 locations are selected in such a way that the distribution in emission can be studied as right-left symmetry, dorsal-ventral symmetry, and the ratio between the central body part and extremities. The body locations (left and right) and their abbreviations are: Forehead (Fhd), Cheek (Chk), Thorax-anterior (Th-A), Thorax-posterior scapulae (Th-P), Abdomen-anterior (Ab-A), Abdomen-posterior kidneys (Ab-P), Elbow-anterior (El-A), Elbow-posterior (El-P), Handpalm (Ha-p), Hand dorsal (Ha-d), Upperleg-anterior (Up-A), Upperleg-posterior (Up-P), Knee (Kne), Hollow of knee (Hkn) and Foot frontal (Fof).

Figure 4 presents, as representative illustration, the intensities of two subjects recorded in summer (July) between at 7-9 p.m. Each 7 cm diameter location is measured for 2 min. The figures show the pattern of a 36 years (Figure 4A) and 61 years (Figure 4B) old male subject, respectively. Analysis of the data illustrated in figure 4 has shown that a difference of more than 1.1 cps between two locations was statistically significant

PHOTON EMISSION FROM HUMAN BODY

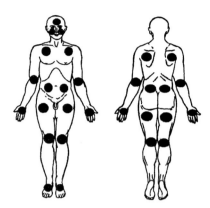

Figure 3. Map of skin locations for measurement of spontaneous photon emission over the anterior (left figure) and posterior (right figure) of the human body.

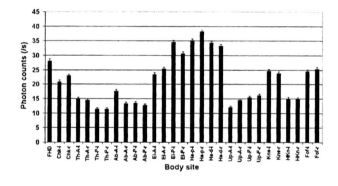

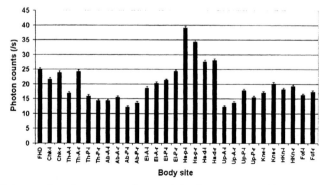

Figure 4. Multi-site recording of spontaneous photon emission from two male subjects: age 36 year (top) and 61 years (bottom).

(p<0.01). It must be noted that the intensities include background noise. This noise is 6 cps; it is largely due to the electronic noise of the photomultiplier tube (about 5 cps). It has been found that subjects have consistently different patterns. The general tendency is that the thorax-abdomen region has the lowest intensity. In contrast the upper extremities and the head region show the highest and most variable levels. The conditions leading to this variability in intensity for corresponding body sites of different subjects are presently unknown. The conditions under study are age, gender, time of day, season, diet, and psychological stress.

5. EFFECTS OF COLORED FILTERS ON SPONTANEOUS VISIBLE EMISSION

Since the hand emits highest spontaneous photon emissions, data for spectral analysis were obtained from a 50x50 mm area of the palm of the hand. We utilized filters with different wavelength cut offs to estimate the spectrum of the emitted light. A square frame of 50 x 50 mm is mounted on a 90 mm diameter ring port extension of the photo-multiplier tube. The distance between the square frame and the port of the photo-multiplier is 80 mm. The recording period was extended to 3 min in order to estimate accurate average values. This was done because the reduced opening at the front of the multiplier tube resulted in photon counts that are low compared to background values.

A set of coloured glass filters manufactured by Scott, is used to facilitate wavelength estimation. Each filter permits wavelength transmission with sharp cut off. The following filters were utilized: WG320, WG360, GG420, GG470, OG530, OG570, RG630 and RG695 nm, with cut-off wavelengths at 320, 360, 420, 470, 530, 570, 630, and 695 nm respectively. The last two filters are at the limit of spectral sensitivity of the photo-multiplier; they are considered internal controls.

The protocol for recording is as follows. Photon emission is recorded in the absence and then with different filters in the sequence from 320 nm to 695 nm, then in the sequence from 695 nm to 320 nm, and no filter. This sequence was repeated once, resulting in 4 data per filter. This protocol for recording was carried out in a period of approximately 2 hours. In this time period the average emission remained quite stable.

The protocol for data analysis was as follows. The data recorded for each filter are combined, and mean and standard error of the mean are calculated. The steps in the calculation procedure are according to Inaba.[16] Results are illustrated in Table 1 and Figure 5. Values are expressed in photon counts per min (cpm).

Even after days of dark adaptation, filters continue to emit very low values of delayed luminescence. This tiny amount must be taken into account when hand photon emission is recorded. Therefore, the average spontaneous luminescence of a filter is subtracted from the photon emission recorded with that respective filter. The difference [(hand + filter) − (filter)] equals the hand emission recorded with each filter. In this example the difference is statistically significant with $p< 0.0000$ for cut-off filters 320, 360, 420, 470 nm, and $p= 0.01$ for 530 nm filter. Hand emission was no longer detectable with cut-off filters at 570 nm and higher.

The statistically significant differences are used for the calculation of the spectrum (Table 1, lower part). The differences in emission intensities between filters with successive cut-off wavelength represent the emission intensity in that particular wavelength sub-range. These intensity values are then combined with the technical specifications of photo-multiplier tube. The EMI 9235QB records only between 200 and 650 nm (ultraviolet, blue, green, yellow, orange and a small fraction of red). The spectral distribution was achieved by computing the average counts for a standardized wavelength range, followed by a mathematical correction for the average sensitivity of recording in each wavelength range.

Figure 5 shows the spectral analysis of the photon emission of the hand as it was recorded with the present protocol and technique. In this example emission is recorded in the 420-570 nm range. It was neither detected in the 200-420 nm nor in the 570-650 nm range. The latter is, at least partially, explained by the combination of low intensity and low spectral sensitivity of the photomultiplier in this range.

Table 1. Data analysis for spectral distribution

Filter cut-off (nm)	Hand+filter (cpm)	Filter alone (cpm)	Hand (cpm)	p
320	1410±19	684±12	726	0.0000
360	1230±16	510±11	720	0.0000
420	1362±18	684±11	678	0.0000
470	1116±14	612±11	504	0.0000
530	552±11	468±12	84	0.01
570	426±11	438±11	-12	ns
630	450±11	432±10	18	ns
695	408±10	420±11	-12	ns

Wavelength range (nm)	Difference between consecutive cut-off filters (cpm)	Average quantum efficiency in wavelength range	Average intensity in wavelength range (cpm)	% intensity in wavelength range (% cpm)
200-320	12	.30	17	0.4
320-360	6	.34	22	0.6
360-420	42	.33	105	2.7
420-470	174	.29	599	15.6
470-530	420	.20	1752	45.6
530-570	96	.09	1342	35.0
570-630	nd	nd	nd	nd
630-695	nd	nd	nd	nd

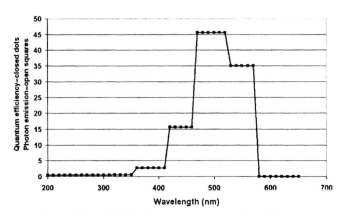

Figure 5. Spectral distribution of spontaneous emission of the hand

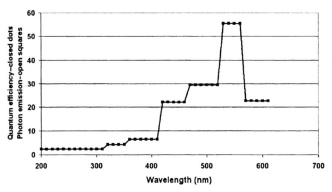

Figure 6. Spectral distribution of light-induced delayed luminescence

Table 2. Spontaneous photon emission of the hand during blood flow restriction with a tourniquet

Thightness of tourniquet	Photon emission before tourniquet (cps)	Photon emission during tourniquet (cps)	Photon emission after tourniquet (cps)
Type 1 experiment: Mildly thight	26.7±.61	24.5±.75	25.7±.58
Type 2 experiment: Thight	26.2±.61	21.8±.68	25.7±.56
Type 3 experiment: Severely thight	30.3±.65	23.4±.68	25.9±.55

6. EFFECTS OF COLORED FILTERS ON LIGHT-INDUCED DELAYED LUMINESCENCE

A comparison between the spectrum of spontaneous and light-induced delayed luminescence of the hand was made. The hand was illuminated using the Ultra Vitalux lamp as described in the methods section. Delayed luminescence was determined by counts over a period of 7-240 s. Figure 6 illustrates the final spectrum. The delayed luminescence was between 420 and 630 nm. Therefore, the induced emission presents a broad band and the spontaneous emission of the same hand has a more limited band.

7. SPONTANEOUS EMISSION AND BLOOD SUPPLY

In an attempt to investigate one physiological aspect of photon emission etiology, we placed a tourniquet around the upper arm to depress the supply of oxygen and nutrients to the hand. Three types of experiments were performed representing increasing degree of tightness (Table 2). The mean photon emission of the hand was significantly decreased more when the tourniquet was more tightened. On removing the tourniquet the photon emission rate was restored in the next five minutes. However, the tighter the tourniquet, the slower was the recovery.

8. DISCUSSION

Although delayed luminescence of human skin has previously been demonstrated [15], its decay, particularly after excitation with strong (artificial) sunlight, has not been studied. We report in this paper that illumination with artificial light that includes a UV component leads to a delayed luminescence with a hyperbolic decay. This characteristic was also observed for natural sunlight. Since light-induced photon emission fades according to hyperbolic law, it takes a long time, and has to be carefully controlled. If not, external conditions can easily interfere with the recording of spontaneous emission. Based on these decay kinetics and experience with long-term dark-adapted skin, we will only record when skin has not been exposed to bright (UV rich) skylight for several hours, and dark-adapted for at least 1 h.

Our protocol for recording skin photon emission bilaterally over several body parts is based on a highly sensitive conventional photomultiplier designed for scanning and movable in three directions. Previously, Cohen and Popp[15] utilizing this device reported that emission from local skin areas is generally stable in a daily experimentation period of several hours. In the present study we have determined more accurately with this device the recording time necessary to obtain reliable average value for the intensity. In this part of the study we observed the large variations in photon count time series, with fluctuations even in the range of tens of seconds. Although the fluctuations are in the range of tens of seconds, reliable average values can be obtained when measurement time is extended to 2-5 min, depending on the size of the opening at the port of the photomultiplier. The origin of such fluctuations is unclear, but they have mainly a biological origin. Results suggest that spontaneous skin visible photon emission fluctuations are due to auto-regulated slow oscillations in local skin blood flow. The tourniquet experiments can be considered as an indication that blood flow (oxygen) is an important factor in the fluctuations of photon emission of the skin. Suppression of photon emission by limiting blood flow has also been reported from cortex of rats; intensity was associated with cerebral blood flow and was depressed after cardiac arrest.[17]

In the present study we demonstrate the capacity of the photo-multiplier tube EMI 9235QB for analysing the spectral distribution of photon emission of the hand within the sensitivity of the photo-multiplier tube (200 – 650 nm). Recording of photon emission was carried out with the equipment for multi-site recording when the photomultiplier port is at a distance of approximately 80 mm from the skin. In multi-site recording a skin area of 9 cm diameter is commonly recorded. Limiting the measured surface actualized the spectral analysis of emissions from the central part of the hand. The smaller dimensions were selected because differences may occur between the central part and the surrounding areas of the palm. Although the size of the measured surface is limited reliable average values can be obtained when measurement time is extended, and repeated measurements are carried out. Furthermore, care has to be taken for the emission of the filter itself, which is unavoidable.

Spectral analysis is commonly utilized to identify the source of photon emission. Limitations are determined by the spectroscopy equipment utilized.[16] The intensity of very weak spontaneous emission cannot be captured with prism dispersion utilizing a narrow monochromator. Interference filters also have severe limitation because the transmission coefficients of these filters are usually small. In addition, the photon emission must be perpendicular. This is usually accomplished by only providing a narrow window to adopt the light to the filter and subsequently to the photo-multiplier. This decreases again the number of photons recorded. Instead, cut-off glass filters with high transmittances are utilized to prevent loss of photons in recording very weak emission. We recorded spontaneous emission with a maximum between cut-off filters of 470 and 570 nm.

Spectral analysis has been the major method applied in order to identify more precisely the source of the photon emission. The maximum for spontaneous emission strongly suggests the participation of reactions involving free radicals and lipid peroxidations, as has been demonstrated for other organs.[18] Comparing the spectral distributions during spontaneous emission and light-induced delayed luminescence indicate a red-shift in the induced delayed luminescence state. A larger fraction of blue green emission was detected in the spontaneous emission state. The difference might be explained by the participation of different reactions in the photon emission. Excited carbonyl compounds (>C=O)* and dimoles O_2^* ($^1\Delta_g$) are produced by dismutation of peroxyradicals and peroxide cleavage. The lipid decomposition step is (2 >CHOO· → C=O + >COH + O_2), in which either the carbonyl group or the oxygen can be formed in electronically excited states. Emission from excited carbonyl groups can then occur at wavelengths, which are in the 400-500 nm range in the case of in vivo sources. In addition, excited singlet oxygen molecules can emit around 780 nm. Other singlet oxygen emission lie even further in the infrared. The excited singlet oxygen molecules may also excite a secondary emission by reacting with unsaturated lipids to produce excited carbonyl groups, which can then emit with wavelengths in the same vicinity (400-500nm). Another possible emission is that arising after dimerization of excited singlet oxygen, the dimer emitting two broad bands centered on 634 and 703 nm. Excited carbonyl groups produced by the first two reactions are held responsible for the blue-green components of the emission, while the emission from excited oxygen dimers is most compatible with the red emission.[18]

9. PERSPECTIVES

Our investigations suggest that the body photon emission measurement is noninvasive with immense potential applications, in particular with respect to the destructive activity of Reactive Oxygen Species (ROS) and their role in the development of chronic diseases. The first multi-author review on the connection between photon emission, stress and disease suggested the implication of oxygen radicals in the pathogenesis of several crippling chronic diseases.[6] Many human and animal studies have demonstrated that in ischemic heart disease, diabetes, rheumatoid arthritis, atherosclerosis, Crohn's Disease, neurological disorders, cancer, ROS and the products of their non-specific interactions with biological molecules are elevated. Recently, it was documented that the intensity and/or symmetry of visible hand emission is disturbed in disease.[19] These studies support the suggestion that, at least visible photon emission from hands provides information on pathophysiological states in a non-invasive manner. It also raises the question which body locations offer optimal information on ROS alterations in stress and disease.

A variety of life situations leading to fight-or-flight response usually increase the level of physiological stress reactions. Such stress can also induce alterations in oxidative status reflected by increased plasma superoxides as well as modified antioxidant defense.[20] It has been commonly accepted that physiological changes following meditation counterbalance high stress levels characteristic of many psychosomatic diseases. The influence of meditation on free radical activity has been studied.[21] The preliminary findings suggest that lower peroxide levels are associated with stress reduction using meditation. Another preliminary study provides initial evidence that human photon emission is influenced by daily stress and meditation.[22] The latter data demonstrated an almost equal photon emission over the body early after sleep. The data also demonstrated an increase in emission as a typical day progresses, as expected by exposure to daily-life stress. Photon emission is reduced during meditation and continues at the low level of intensity for some time after meditation has stopped.

In summary, we hope that the present studies regarding human photon emission will contribute to future technological developments in smaller, versatile and affordable devices. In concert with low cost, meditation-type interventions, this holds a promise for a broadly applicable innovative strategy, targeting both stress related and chronic degenerative ailments.

10. ACKNOWLEDGEMENTS

This work was supported by the Fred Foundation, the Samueli Institute of Information Biology, and the Foundation for Bioregulation Research. The authors thank Fritz-Albert Popp and Yu Yan for their supportive and critical suggestions that have helped us in analysing and refining these complex experiments. The authors also thank John Ackerman for his comments on earlier versions and editing of the text, considerably improving it's readability.

11. REFERENCES

1. V. V. Perelygin and B. N. Tarusov, Flash of very weak radiation on damage to living tissues, *Biophysics* **11**, 616 – 618 (1966).
2. V. V. Artem'ey, A. S. Goldobin and L. N. Gus'kov, Recording the optical emission of a nerve, *Biophysics* **12**, 1278 – 1280 (1967).
3. V. V. Blokha, G. V. Kossova, A. D. Sizov, V. A. Fedin, Y. P. Kozlov, O. R. Kol's, and B. N. Tarusov, Detection of the ultraweak glow of muscles on stimulation, *Biophysics* **13**, 1084 – 1085 (1968).
4. I. G. Shtrankfel'd, L. L. Klimensko, and N. N. Komarow, Very weak luminescence of muscles, *Biophysics* **13**, 1082 – 1084 (1968).
5. F. A. Popp, A. A. Gurwitsch, H. Inaba, J. Slawinski, G. Cilento, F. A. Popp, K. H. Li, W. P. Mei, M. Galle, R. Neurohr, R. Van Wijk, D. H. J. Schamhart, W. B. Chwirot, and W. Nagl, Biophoton emission, Multi author review, *Experientia* **44**, 543 – 600 (1988).
6. R. Van Wijk, R. N. Tilbury, J. Slawinski, A. Ezzahir, M. Godlewski, T. Kwiecinska, Z. Rajfur, D. Sitko, D. Wierzuchowska, B. Kochel, Q. Gu, F. A. Popp, E. M. Lilius, P. Marnila, R. Van Wijk, and J. M. Van Aken, Biophoton emission, stress and disease, Multi-author review, *Experientia* **48**, 1029-1102 (1992).
7. E. Cadenas, I. D. Arad, A. Boveris, A. B. Fisher, and B. Chance, Partial spectral analysis of the hydroperoxide-induced chemiluminescence of the perfused lung, *FEBS Lett.* **111**, 413-418 (1980).
8. H. Inaba, Applications of measuring techniques of extremely weak light to medicine and life sciences, *Kogaku (Optics)* **12**, 166-179 (1983).
9. E. Cadenas, A. Boveris, and B. Chance, Low-level chemiluminescence of bovine heart submitochondrial particles, *Biochem. J.* **186**, 659-667 (1980).
10. E. Cadenas and H. Sies, Low level chemiluminescence of liver microsomal fractions initiated by tert-butyl hydroperoxide. Relation to microsomal hemoproteins, oxygen dependence and lipid peroxidation, *Eur. J. Biochem.* **124**, 349-356 (1982).
11. M. Nakano, K. Takayama, Y. Shimizu, Y. Tsuji, H. Inaba, and T. Migita, Spectroscopic evidence for the generation of singlet oxygen in self-reaction of sec-peroxy radicals, *J. Am. chem. Soc.* **98**, 1874-1975 (1976).
12. R. Edwards, M. C. Ibison, J. Jessel-Kenyon, and R. B. Taylor, Light emission from the human body, *Complementary Medical Research* **3**, 16-19 (1989).
13. M. Usa, B. Devaraj, M. Kobayashi, M. Kakeda, H. Ito, M. Jin, and H. Inaba, Detection and characterization of ultraweak biophotons from life processes, in: *Optical Methods in Biomedical and Environmental Sciences*, H. Ohzu and S. Komatsu eds., Elsevier Science Publishers, Amsterdam, pp. 3 – 6 (1994).
14. H. Inaba, Photonic sensing technology is opening new frontiers in biophotonics, *Optical Review* **4**, 1-10 (1997).
15. S. Cohen and F. A. Popp, Low-level luminescence of the human skin, *Skin Research and Technology* **3**, 177-180 (1997).
16. H. Inaba, Super-high sensitivity systems for detection and spectral analysis of ultraweak photon emission from biological cells and tissues, *Experientia* **44**, 550-559 (1988).
17. M. Kobayashi, M. Takeda, K. I. Ito, H. Kato, and H. Inaba, Two-dimensional photon counting imaging and spatiotemporal characterization of ultraweak photon emission from a rat's brain in vivo, *J. Neuroscience Methods* **93**, 163-168 (1999).
18. R. Van Wijk and D. H. J. Schamhart, Regulatory aspects of low intensity photon emission, *Experentia* **44**, 586-593 (1988).
19. S. Cohen and F. A. Popp, Biophoton emission of human body, *Indian J. Exp. Biol.* **41**, 440-445 (2003).
20. I. Cernak, V. Savic, J. Kotur, V. Prokiv, B. Kuljic, D. Grbovic, and M. Veljovic, Alterations in magnesium and oxidative status during chronic emotional stress, *Magnesium Research* **13**, 29-36 (2000).
21. R. H. Schneider, S. I. Nidich, J. W. Salerno, H. M. Sharma, C. E. Robinson, R. J. Nidich, and C. N. Alexander, Lower lipid peroxide levels in practitions of the transcendental meditation program, *Psychosom Med* **60**, 38-41 (1999).
22. E. P. A. Van Wijk and R. Van Wijk, Effect of meditation on photon emission from the hand, *submitted for publication*, (2003).

Chapter 14

LASER-ULTRAVIOLET-A INDUCED BIOPHOTONIC EMISSION IN CULTURED MAMMALIAN CELLS

Hugo J. Niggli[1], Salvatore Tudisco[2], Giuseppe Privitera[2], Lee Ann Applegate[2], Agata Scordino[2], and Franco Musumeci[2]

1. INTRODUCTION

1.1. Overview

Photons or quanta, the smallest elements of light, necessarily play an important role in many atomic and molecular interactions and changes in our physical universe. 1900 Max Planck postulated the quantum elements of light as discrete parcels. Plancks quantum theory was first noted, when Einstein in 1905 used this idea in order to explain the photoelectric effect. On the basis of modern quantum physics it became clear that elementary particles no longer represent separate, but complementary terms. Photons can be regarded as the information carriers of matter because they not only glue elementary particles together but can be exchanged within atomic and molecular interactions.

1.2. History

At the beginning of this century, Alexander Gurwitsch, born 130 years ago in Russia, suggested that ultraweak photons transmit information in cells as summarized previously[1]. Although the results of Gurwitsch were refuted by Hollaender and Klaus[2], the presence of biological radiation was re-examined with the development of photomultiplier tubes in the mid-1950s by Facchini and co-workers[3]. In western countries several pioneers, Quickenden in Australia[4], Popp in Germany[5] and Inaba in Japan[6], independently developed methods for ultraweak photon measurements in a variety of different cells by

---
[1] BioFoton AG, rte. d'Essert 27, CH-1733 Treyvaux, Switzerland, Phone/Fax: +41-26-4131445, E-mail: biofoton@swissonline.ch (corresponding author). Lee Ann Applegate, Department of Obstetrics, Laboratory of Oxidative Stress and Ageing, University Hospital, CHUV, CH-1011 Lausanne, Switzerland.
[2] Laboratori Nazionali del Sud, I.N.F.N. and Dipartimento di Metologie Fisiche e Chimiche per l'Ingegneria, University of Catania, I-95129 Catania, Italy.

the use of an extremely low noise, highly sensitive photon counting system which allows maximal exploitation of the potential capabilities of a photomultiplier tube. This research showed that plant, animal and human cells emit ultraweak photons often called biophotons[1,7-9].

1.3. Biophotonic sources

Tilbury discussed in the multi-author review of Van Wijk[10] that ultraweak photon emission has been detected in both the visible and ultraviolet region. Radiation in the visible region appears to be due to excited carbonyl groups and/or excited singlet oxygen dimers arising from lipid peroxidation, which in turn are associated with an increase in various reactive oxygen species such as superoxide anion, hydrogen peroxide, hydroxyl radical and singlet oxygen. There is also substantial evidence for DNA playing a key role in these emissions[11,12].

1.4. Ultraweak photons in cultured cells

Experiments with cultured human cells were reported[1] in which normal and DNA excision repair deficient Xeroderma pigmentosum (XP) cells were UV-irradiated in medium and balanced salt solution (EBSS) and assessed for ultraweak photon emission. These investigations showed that an important difference between normal and XP cells was present and that XP cells are unable to store ultraweak photons which are efficiently absorbed by normal cells.

Since Hayflick's pioneering work in the early 1960s, human diploid fibroblasts have become a widely accepted *in vitro* model system. Recently, Bayreuther and co-workers extended this experimental approach showing that fibroblasts in culture resemble, in their design, the hemopoietic stem-cell differentiation system[13]. His group reported morphological and biochemical evidence for the fibroblast stem-cell differentiation system in vitro and showed that normal human skin fibroblasts in culture spontaneously differentiate along the cell lineage of mitotic fibroblasts (MF) and post-mitotic fibroblasts (PMF). Additionally, they developed methods to shorten the transition period and to increase the frequency of distinct post-mitotic cell types using physical or chemical agents such as mytomycin C. Mitomycin C is an effective chemotherapeutic agent for several cancers in man; this effect is probably related to the interaction with DNA leading to DNA-DNA crosslinks and DNA-protein crosslinks. We have previously demonstrated that mitomycin C-treatment of three different normal human fibroblast strains (CRL 1221, GM 38 and GM 1717), frequently used in mutation, transformation, and aging research, induces characteristic morphological changes in the fibroblasts and brings about specific shifts in the [^{35}S]methionine polypeptide pattern of total cellular proteins. These results support the notion that mitomycin C accelerates the differentiation pathway from mitotic to postmitotic fibroblasts. Using this system, we were also able to demonstrate that no significant difference exists in the rate and the extent of the excision-repair response to thymine-containing pyrimidine dimers following UV-irradiation shortly after mitomycin C treatment of distinct strains of human skin fibroblasts and in the mitomycin C-induced PMF stage of these cells[14]. In addition, aphidicolin inhibits excision repair of UV-induced pyrimidine photodimers in low serum cultures of mitotic and mitomycin C-induced postmitotic fibroblasts of human skin[15]. Since that fibroblasts play an essential role in skin aging, and wound healing, our results imply that the fibroblast differentiation system is a very useful tool to unravel the complex mechanism of skin aging, skin carcinogenesis and wound healing. For example, using this system we have shown that bone growth factors induce proliferation in fibroblastic differentiation and change the emission of ultraweak photons[1].

Based on all these findings, we describe in this report a highly sensitive technique for UVA-laser-induced ultraweak photon emission in cultured mammalian cells. This new biological model system may open new dimensions on the importance of light in cell biology.

2. MATERIAL AND METHODS

2.1. Cell Culturing

Skin fibroblasts GM 38 (p 9) from a 9 year old black girl were obtained from the Human Genetic Mutant Cell Repository (Camden, NJ, USA). Cloudman S91 mouse melanoma cells of clone m-3 (CCL 53.1; p 40) were purchased from the American Type culture collection (Rockeville, USA). Skin fibroblasts from a Xeroderma Pigmentosum from a 9 year old patient of complementation group A (XPA), XP12BE (GM05509A) were obtained from the Human Genetic Mutant Cell Repository (Camden, NJ, USA). XPA cells derived from a 10-year-old female (CRL 1223) were purchased from the American Type culture collection (Rockeville, USA). Normal cells were plated 1:2, XPA cells 1:3 and melanoma cells 1:10. For the determination of ultraweak photon emission, cells were cultured in tissue culture plastic flasks (surface, 75cm^2, Gibco Basel, Switzerland) in 15 ml Dulbecco's modified Eagle's medium (DMEM; Gibco, Basel Switzerland) supplemented with 10% fetal calf serum and 100 units (U) ml^{-1} penicillin-streptomycin as previously described[1]. Postmitotic fibroblasts were prepared as previously described[15]. Cells were frozen gently in liquid nitrogen in DMEM with 10% DMSO by using the cryopreservation apparatus from Biotech Research Laboratories (Rockeville, MD, USA). Controlled gradual temperature reduction during cryopreservation was critical for the maintenance of cell life and viability. This apparatus preserves cells at the rate of 1°C per minute when placed in a -70° C freezer. After 5 hours or overnight, frozen samples were transferred to the liquid nitrogen storage system. Cells were counted in triplicates (± 10%) in a haemocytometer from Neubauer (Flow Laboratories, Baar, Switzerland).

2.2. Classical Biophotonic Measurements in Human Skin Cells

For ultraweak photon emission measurements different mammalian cell samples were transported in liquid nitrogen. The cells were thawed and were diluted in a volume of 10 ml Dulbecco's modified Eagle's medium, from which phenol red was omitted, and were transferred to a quartz sample glass (2.2x2.2x3.8 cm^3; thickness 0.15 cm). Detection and registration of spontaneous and light-induced emitted photons was accomplished as described before[1]. The test samples were kept in a dark chamber in front of a single photon counting device equipped with an EMI 9558 QA photomultiplier tube (diameter of the cathode 48mm, cooled to -25° C). This high-sensitivity photon counting device is described in detail by Popp and co-workers[5] and measures photon intensities as low as 10^{-17} W in the range between 220 and 850 nm. Signal amplification is normally 10^6-10^7 and dark count ranges between 10-15 counts per second (cps). Maximum efficiency of the S20-cathode is 20-30% at 200-350nm, decaying almost linearly down to 0% at a wavelength of 870 nm and mean quantum efficiency in the entire spectral range is about 10%. The integral intensity values within each interval of 40 ms were stored and processed by an interfaced computer. For light induced emission experiments, irradiation of the test sample was performed perpendicular to the detecting direction with focused white light from a 75-W Xenon lamp. Each measuring cycle was started by irradiating the sample for 5s. The measurements for the monochromatic light induction were performed by changing the spectrum of the inducing light through a monochromomator (PTI, Hamburg, Germany) from 300 to 450 nm (25nm interval).

2.3. Delayed Luminescence Measurements in Mammalian Cells after UVA laser induction

For the measurements of the delayed low level luminescence in cells, they were transported in liquid nitrogen to Catania and diluted in phosphate buffered saline (PBS). The measurements have been performed using the device ARETUSA[16], developed at the National Laboratory of South in Catania. The detection system is shown in Figure 1. This measuring system consists of a cooled single photon count photomultiplier, a highly efficient optical system able to collect the emitted light from the sample and to transmit it to the photomultiplier, a high intensity pulsed nitrogen laser with pulses at 337 nm, an electronic device able to deactivate the photomultiplier during the laser pulses and an electronic set-up able to coordinate the measurement procedure and to collect, analyse and elaborate the signals from the photomultiplier. For the photomultiplier the Hamamatsu R1878 was used. It is a multialkali photocathode having a spectral response from 300 nm up to 850 nm with a maximum response at 420 nm. Its effective area is small (only 4mm in diameter) and therefore a high optical efficiency of the set up is required. However, under these conditions its background noise is lowered to 0.01 compared to that of 2 inches photomultiplier as described previously[7, 17]. In order to further reduce the noise the photomultiplier was cooled down to -20°C using an original forced circulation cooling system which is placed to direct contact with the lateral surface of the photomultiplier and 100 µl of cell suspension was placed directly, without using any cuvette, on the upper quartz termination of a liquid light guide.

As light source, we used a pulsed nitrogen laser (LASERPHOTONIC LN203C, Catania, Italy) characterised by a wave-length of 337.1 nm, a pulse width of 5 ns and an energy of about 100 µJ/pulse (+ 3% standard deviation). In order to receive a uniform illumination of the measured samples (an hemisphere having a diameter of approximately 5 mm) a lighting system based on a trifurcated quartz fiber was used. The three terminations surrounds the sample at a constant distance with an angle of 120° each other. Under these conditions almost all of the light emitted by the laser source reach the sample with a very high power per unitary volume (about $6*10^{11}$ W/m^3).

In the system presented here, an electronic gate was developed, controlled by a digital signal, able to establish on all dynodes, if digital signal have his lower level, the same tension as a normal voltage divider circuits, and, if digital signal have his higher

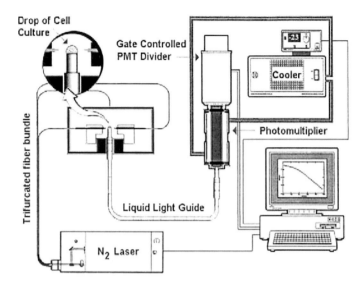

Figure 1. A block diagram of the highly sensitive UVA-laser induced ultraweak photon emission detection system.

level, to invert the sign of the voltage between the cathode and the first dynode and between the third and the second dynode. Using this protocol, it allows us to start the photon counting few microseconds after the laser pulses have reached the sample.

The measurement process is performed and controlled by a microcomputer boosted through a Ortec Multi-channel Scaler (MCS) plug-in card characterised by dwell times ranging from 100 ns to 1300 s, a memory length of 65.536 channels, and input counting rates up to 150 MHz. The signals comings from PMT are first processed by an octal discriminator (Lecroy 6408B) and then the TTL output is send to the MCS for acquisition. The synchronization of the process is managed by MCS. Its start-out TTL signal is split in three identical signals each of them addressed to a gate and delay generator (Ortec GG 8000) in order protect temporary the PMT during the sample irradiation. After a delay time of about 6 µs the inhibition stops and UVA-laser-induced delayed luminescence will be measured.

3. RESULTS AND DISCUSSION

3.1. Biophotonic emission in cultured cells with the classical design

Using the classical design for ultraweak photon measurements, we have found in repair deficient Xeroderma pigmentosum cells hyperbolic re-emission curves as depicted in Figure 2. These results confirm several previous investigations in plant and mammalian

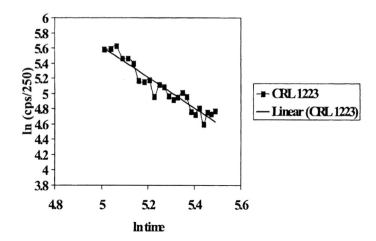

Figure 2: White light-induced ultraweak photon emission dynamic of 10^7 repair deficient XPA skin fibroblasts (CRL 1223) in 10 ml suspension. One time interval correspondens to 4 ms. First measurment was after 150 ms. Experimental data (■) compared to hyperbolic decay by a theoretical calculated line. Correlation coefficient calculated by liner regression is 0.96.

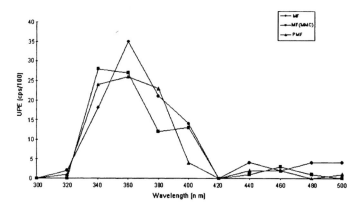

Figure 3: Initial counts of photoinduced ultraweak photon emission (UPE) registered in 10ms after illumination with monochromatic light during 10 s in distinct stages of XPA human skin fibroblasts (CRL1223; 3×10^6 cells/10ml suspension solution). Average of two independent experiments. Standard deviation is $\pm 10\%$.

cells[1,5,12]. As interpreted by Popp and co-workers[18], the hyperbolic decay kinetics found in living systems after pre-illumination with white light may indicate coherent re-scattering of ultraweak photons due to collective excitation of nucleic acids within the DNA of the investigated fibroblasts.

As described above, Bayreuther and co-workers[13-15] showed biochemical and morphological evidence for the fibroblast differentiation system in vitro. They showed that normal human skin fibroblasts in culture spontaneously differentiate along the cell lineage of mitotic (MF) and postmitotic fibroblasts (PMF). Additionally, they developed methods to shorten the transition period and to increase the frequency of distinct postmitotic cell types using physical agents such as ultraviolet light (UV) and mitomycin C (MMC).

Figure 3 depicts initial light-induced photon emission in distinct differentiation stages of XPA fibroblasts (CRL1223) after monochromatic induction between 300-500 nm. There is no discernible difference between untreated mitotic fibroblasts (MF), MMC treated mitotic cells and MMC-induced postmitotic cultures (PMF) one weeks following MMC-treatment. Similiar results (not shown) were obtained for a normal cell line (GM 38).

Our results confirm our previously reported finding that the most important induction range for these very weak photons is the UVA range between 330-380nm[1]. Furthermore, they support our results of unchanged white-light-induced ultraweak photon emission in different stages of fibroblastic differentiation shown in the same report[1].

3.2. Ultraweak photon emission in mammalian cells following irradiation with a nitrogen laser in the UVA-range

Based on the finding that the most important induction range for these very weak photon emission ist in the the UVA range, we developed a highly sensitive technique for UVA-laser-induced ultraweak photon emission. The system used was a UVA nitrogen laser characterised by a wavelength of 337 nm. The laser-pulse of 5 ns allowed electronically closing of the photomultiplier system. As a consequence of this change the initial count measurement could be performed after less than 10µs instead of 150 ms as shown in Figure 2 above. Figure 4 shows a typical experiment of Cloudman S91 mouse melanoma cells at a cell density of 1 million cells/ml.

It is important to note in this Figure 4 that our new sophisticated detection procedure allows us to determine this light induced ultraweak photon emission in a very short time interval of µs instead of ms as depicted in Figure 2 using our classical design. This important change leads to a light burst of more than 10^7 photons compared to several hundreds as shown in the same Figure 4.

Figure 4 confirms a further remarkable feature of delayed ultraweak luminescence which has been recently described[18]: the fact that cells display sinusoidal oscillations around a hyperbolic decay function. It should be noted that this relaxation cannot be assigned to the usual optical transition of isolated- or triplet states as found for example for the picosecond fluorescence decay time of calf thymus DNA reported by Georghiou and co-workers[20]. This follows from the rather long decay time, lasting at least several hundred milliseconds as shown in Figure 2.

Initial white light induced ultraweak photon emission after 150 ms (dwell time 50ms) in repair deficient XPA-monolayer fibroblasts (GM5509) is compared to initial values

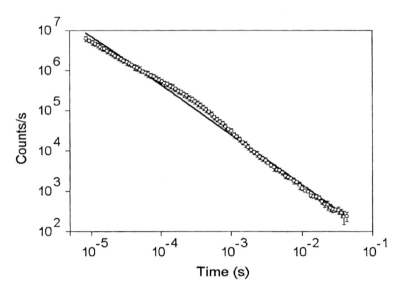

Figure 4: UVA laser light-induced ultraweak photon emission dynamic of 10^6 mouse melanoma cells (Cloudman S91) in 100 μl PBS suspension. One time interval correspondens to 1 μs during the first nine measurements and was gradually increased up to 5 ms for the final point. First measurment was after 8.5 μs. Experimental data (O) compared to hyperbolic decay by a theoretical calculated line. Correlation coefficient calculated by liner regression is 0.999. Each point represent average of 100 determinations.

following UVA-laser induction in the same fibroblasts after 8.5 μs (dwell time 1 μs). With 10^6 cells/ml we obtained with the classical design 766±20 counts in contrast to 3.4±0.04 million counts using the new determination system. Overall, the improvement factor for the detection of ultraweak photons in fibroblasts using the classical design compared to the new UVA-laser induction procedure is about 10^4. It has to be noted that our classical device uses a mechanically based closing shutter system in the time range higher than 100 ms and cell quantities up to 10 ml. Our new detection procedure is able to measure ultraweak photon emission in cell quantities as low as 100 μl which is a factor of 100 lower compared to the previous procedure (10ml). Therefore the total improvement with our new design is increased by a factor of up to 10^6.

3.3. Conclusion

In conclusion, a new highly sensitive method using UVA-laser induced delayed ultrawek luminescence in cultured mammalian cells is presented. Our results show evidence that this biological system is a new powerful non-invasive tool to determine biophysical changes within the cells. Future research will show whether this new biological detection system can open new scientific strategies in the field of cell biology.

4. ACKNOWLEDGEMENTS

This study was funded in part by grants from the Swiss League Against Cancer (KFS 695-7-1998) and the Erwin Braun Foundation. HN was generously supported by the H +M Kunz Stiftung (Urdorf, Switzerland) as well as by his parents Alfred and Emmy Wehrli (Aarau, Switzerland).

5. REFERENCES

1. H. J. Niggli, C. Scaletta, Y. Yan, F. A. Popp and L. A. Applegate, Ultraweak photon emission in assessing bone growth factor efficiency using fibroblastic differentiation, *J. Photochem. Photobiol. B: Biol.* **64**, 62-68 (2001).
2. A. Hollaender and W. Klaus, An experimental study of the problem of mitogenetic radiation, *Bull. Nat. Res. Council* **100**, 3-96 (1937).
3. L. Colli, U. Facchini, G. Guidotti, R. Dugnani Lonati, M. Arsenigo and O. Sommariva, Further measurements on the bioluminesence of the seedlings, *Experientia* **11**, 479-481 (1955).
4. T. I. Quickenden and S. S. Que-Hee, The spectral distribution of the luminescence emitted during growth of the yeast Saccharomyces cerevisiae and its relationship to mitogenetic radiation, *Photochem. Photobiol.* **23**, 201-204 (1976).
5. F. A. Popp and L. Beloussov, *Integrative Biophysics: Biophotons* (Kluwer Academic Publishers, Dordrecht, 2003).
6. H. Inaba, Y. Shimizu, Y. Tsuji and A. Yamagishi, Photon counting spectral analyzing system of extra-weak chemi- and bioluminescence for biochemical applications, *Photochem. Photobiol.* **30**, 169-175 (1979).
7. A. Scordino, A. Triglia, F. Musumeci, F. Grasso and Z. Raifur, Influence of the prescence of Atrazine in water on in-vivo delayed luminescence of acetabularium acetabulum, *J. Photochem. Photobiol. B: Biol.* **32**, 11-17 (1996).
8. R. Van Wijk and H. Van Aken, Spontaneous and light-induced photon emission by rat and by hepatoma cells, *Cell Biophys.* **18**, 15-29 (1991).
9. F. Grasso, C. Grillo, F. Musumeci, A. Triglia, G. Rodolico, F. Cammisuli, C. Rinzivillo, G. Fragati, A. Santuccio and M. Rodolic, Photon emission from normal and tumor human tissues, *Experientia* **48**, 10-13 (1992).
10. W. B. Chwirot, G. Cilento, A. A. Gurwitsch, H. Inaba, W. Nagl, F. A. Popp, K. H. Li, W. P. Mei, M. Galle, R. Neurohr, J. Slawinski, R. V. Van Wijk and D. H. J. Schamhart, Multi-author review on Biophoton emission, *Experientia* **44**, 543-600 (1988).
11. B. Devaraj, R. Q. Scott, P. Roschger and H. Inaba, Ultraweak light emission from rat liver nuclei, *Phochem. Photobiol.* **54**, 289-293 (1991).
12. H. J. Niggli, The cell nucleus of cultured melanoma cells as a source of ultraweak photon emission, *Naturwissenschaften* **83**, 41-44 (1996).
13. K. Bayreuther, H. P. Rodemann, R. Hommel, K. Dittman, M. Albiez and P. I. Francz, Human skin fibroblasts in vitro differentiate along a terminal cell lineage, *Proc. Natl. Acad. Sci., USA* **85**, 5112-1516 (1988).
14. H. J. Niggli, K. Bayreuther, H. P. Rodemann, R. Röthlisberger and P. I. Francz, Mitomycin C-induced postmitotic fibroblasts retain the capacity to repair pyrimidine photodimers formed after UV-irradiation. *Mutation Res.* **219**, 231-240 (1989).
15. H. J. Niggli, Aphidicolin inhibits excision repair of UV-induced pyrimidine photodimers in low serum cultures of mitotic and mitomycin C-induced postmitotic human skin fibroblasts. *Mutation Res.* **295**, 125-133 (1993).
16. S. Tudisco, G. Privitera, A. Scordino and F. A Musumeci, A New Advanced Research Equipment for Fast Ultraweak Luminescence Analysis – ARETUSA, in: *Proceedings of the international Wokshop "Energy and information transfer in biological systems: How physics could enrich biological understanding"*, edited by S. Tecla, F. Musumeci, L. Brizik, M.W. Ho (World Scientific, Singapore, 2003) pp. 308-318.
17. R. Van Wijk, A. Scordino, A. Trigla and F. Musumeci, Simulataneous measurements of delayed luminescence and chloroplast organization from Acetabularia acetabulum,. *J. Photochem. Photobiol. B: Biol.* **49**, 142-149 (1999).

18. F. A. Popp and Y. Yan, Delayed luminescence of biological systems in term of coherent states, *Physics letters A* **293**, 93-97 (2002).
19. F. A. Popp, J. J. Chang, A. Herzog, Z. Yan and Y. Yan Evidence of non-classical (squeezed) light in biological systems, *Physics letters A* **293**, 98-102 (2002).
20. G. Ge and S. Georghiou, Excited-state properties of the alternating poly (dA-dT)· poly(dA-dT). *Photochem. Photobiol.* **54**, 301-305 (1991).

Chapter 15

BIOPHOTON EMISSION AND DELAYED LUMINESCENCE OF PLANTS

Yu Yan[1]

1. INTRODUCTION

Spontaneous biophoton emission (BPE) and delayed luminescence (DL) of plants have been studied since five decades [1-17]. In 1954, Colli and Facchini discovered BPE in germinating plants [1]. Since then, BPE has been observed in plant tissues, cells [3-8] and isolated chloroplasts [9]. The BPE is considered to originate from oxidative metabolic processes [6-9] and photon storage of macromolecules [3-5]. DL was discovered in green plants first by Strehler and Arnold 1951 [2]. After illumination with red and infra-red light, a relative maximum appears in the decay of DL emitted by plants [10]. The most popular explanation for the origin of DL traces it back to light emission of charge recombination during a back-flow of charges in the photosynthetic electron-transport chain PS II and PS I [11-12]. The relative maximum occurs because of the addition of different decay curves of photon emission from PS II and PS I [11] or from different sub-populations of PS II [12]. DL of plants shows hyperbolic relaxation and follows the same kinetics at different wavelengths in the visible range [13-15, 17]. This observation leads to the explanation of origin of DL in terms of coherent states. Popp et al. proposed that DL may reflect exited states of a coherent biophoton field in each living system [14-15]. In this case, the relative maximum is a part of the oscillation of the DL decay curve [14-16]. This oscillation can be understood as a consequence of the coupling of at least two coherent states [15]. Also based on the idea of coherent states, Musumeci et al. postulated that the DL originates from coherent solid states in cells, e.g. high-ordered cytoskeleton [13].

In the present work, the BPE of germinating barley seeds and the DL of green leaves, leaf homogenates and isolated chloroplasts were investigated. The fit of the DL relaxation curves to the hyperbolic and oscillation functions predicted by the theory of coherent states was examined. The applications of BPE and DL in the research of plants were discussed.

---

[1] International Institute of Biophysics, Neuss, Germany, www.lifescientist.de, E-mail: yanyu@lifescientists.de

2. MATERIALS AND METHODS

2.1. Instrument of BPE and DL measurements

All BPE and DL measurements were performed in the Photon-Measurement-System-2 (PMS-2) depicted in Figure 1, which was developed by our own research group. This system is equipped with two photomultipliers (Thorn Emi, type 9558 QA) as photon detectors. For the purpose of DL measurement, one of the three light sources was employed in order to illuminate a leaf sample: an Argon-laser (Melles Griot, type 35MAP431-230, 457nm, 12mW), a stabilized He-Ne-laser (Melles Griot, type 05STP903, 632.8nm, 1mW) or five infrared LEDs (Roithner Lasertechnik, type ELD-780-514, 780nm, 11mW each).

2.2. Measurement of BPE of germinating barley seeds

After removal of seed coats, 10 barley (*Hordeum vulgare L.*) seeds were put in a quartz cell ($22 \times 22 \times 38$ mm^3, Hellma, type SUPRASIL) in such a way, that the embryos were at the upside. 1ml distilled water was added in the cell, so that only the lower half of a seed was under water. A quartz plate covered the cell in order to avoid loss of water. The covered quartz cell was then put in the dark chamber of PMS-2 and measured continuously in the next 6 days. In a parallel experiment, 10 barley seeds germinated under the same condition, and photographs of the seeds or seedlings were taken at different stages.

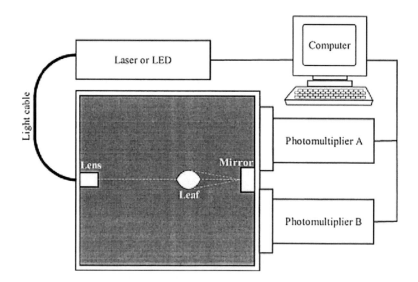

Figure 1. Photon-measurement-system-2 (PMS-2)

2.3. Measurement of DL of plant leaves

Subjects of investigation were leaves of three kinds of indoor plant, i.e. *Crassula ovata*, *Anthurium* and *Kalanchoe daigremontiana*, and leaves of corn salads (*Valeriana locusta*). The homogenate of corn salad leaves prepared in 0.4M sucrose buffer solution containing 60mM K_2HPO_4, 30mM KH_2PO_4, 5mM $MgCl_2$ and 35mM NaCl was also investigated. Furthermore, the chloroplasts were isolated from the homogenate through sucrose gradient centrifugation described by Richter[19] and investigated. In each measurement, a whole leaf, 10ml homogenate or 10ml solution containing isolated chloroplasts was placed in the dark chamber of PMS-2, illuminated 30s by the LED or laser, and measured afterwards by the photomultipliers.

3. RESULTS AND DISCUSSION

3.1. BPE of germinating barley seeds

One typical curve of BPE-dynamics during germination is shown in Figure 2a. Other curves presented in Yan's dissertation show similar patterns [20]. The BPE is high immediately after water addition. In the first 6 hours, the BPE-intensity decreases rapidly. The decrease of BPE-intensity slows down in the next 6 hours and then turns to increase (Figure 2a arrow 1). From the 20th hour on, BPE-intensity begins to increase rapidly till about 100 hours after imbibition (Figure 2a arrow 4). The increase rate shows a rhythmic variation with a period of about one day (1200min to 1600min, Figure 2a arrow 2, 3, 4). After that, the BPE-intensity enters a stagnation phase. The imbibed dead seeds show only the beginning decrease of BPE-intensity and no later increase (Figure 2b).

The photographs show four important time points in the germination procedure of barley seeds (Figure 3). The barley seeds are dry at the beginning (Figure 3a) and fully soaked after 11 hours (Figure 3b). In this period, the BPE-intensity decreases and reaches the minimum while the seeds are fully soaked. From the 21st hour on, coleoptiles and radicles begin to penetrate (Figure 3c) and grow up (Figure 3d). The BPE-intensity enters a rapid increasing phase during the growth of seedlings.

CO_2 shows a strong inhibitory effect on the BPE of barley seedlings (Figure 4). A CO_2-current or an air-current (2.7 l/min) was blown in the dark chamber (volume: 23 l) during the BPE measurement. The BPE-intensity decreases immediately after the CO_2-treatment and increases by the air ventilation. Although CO_2 inhibits a major part of BPE, the BPE of barley seedlings doesn't go down to background (10-15 counts/5s), especially when the seedlings have already been treated by CO_2 several times.

A germination process can be divided into three phases according to the speed of water uptake [18]. In the first phase, a dry seed soaks up water rapidly. This phase is a physical procedure that takes place in a living seed as well as in a dead seed. In the case of barley seeds it takes 6 to 8 hours [21]. In the second phase, the water uptake stagnates. Water is re-distributed in the seed. Proteins are synthesized by use of existing mRNA. Mitochondria are re-organized and begin to function. In the third phase, shoots penetrate out of the seed, and the water uptake starts again.

The BPE dynamics correlate well to the triphasic germination procedure. In the first phase, the BPE-intensity decreases fast. This decrease takes place not only by imbibed living seeds but also by imbibed dead seeds. It shows the physical property of this phase.

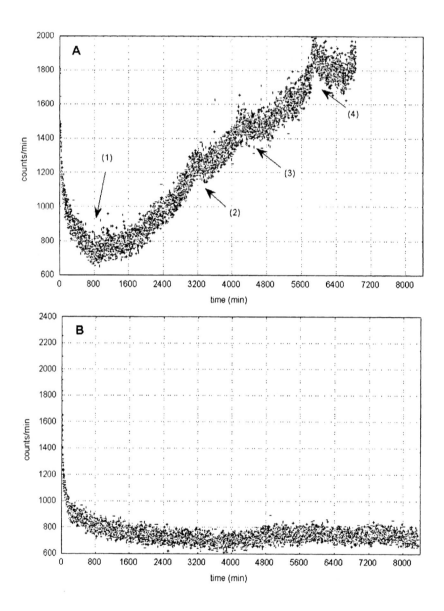

Figure 2. BPE-dynamics of 10 living barley seeds (A) and 10 dead barley seeds (B) after water addition.

(a) 10 barley seeds in distilled water, 0 hour
(b) 10 barley seeds in distilled water, 11 hours
(c) 10 barley seeds in distilled water, 21 hours
(d) 10 barley seeds in distilled water, 74 hours

Figure 3. Photographs of germinating barley seeds in four stages.

The BPE-intensity turns around in the second germination phase. Despite the high biochemical activities in this phase, the BPE-intensity is at the lowest level. From the 21st hour on, the germination procedure reaches its third phase, and coleoptiles and radicles begin to grow. The BPE-intensity begins to increase too. In addition, the periodic variation of the increase rate of BPE-intensity reveals a circadian rhythm in the growth of seedlings. It shows that even in darkness the growth of plant tissues has a daily rhythm.

The strong inhibitory effect of CO_2 on BPE means that respiration plays an important role in BPE of barley seedlings. However, there is still a minor part of BPE, which cannot be inhibited by CO_2. This part of BPE has to be investigated in further studies.

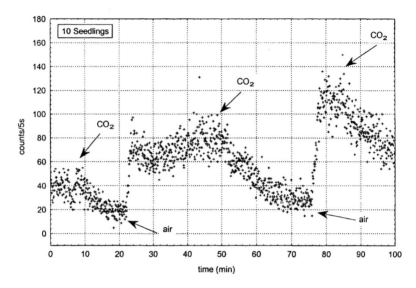

Figure 4. BPE of barley seedlings while the dark chamber was blown by CO2 or by normal air.

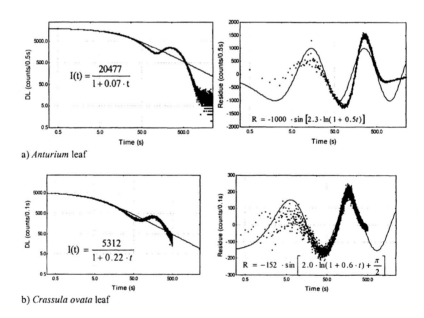

a) *Anturium* leaf

b) *Crassula ovata* leaf

Figure 5. Two DL decay curves of leaves from two plant species.

3.2. DL of plant leaves

After an infrared or red light illumination, the DL of a whole leaf lasts over 10 minutes. The decay curve of DL shows an oscillation around a hyperbolic function. The residues from the fitting to a hyperbolic function can be well fitted to an oscillation function (Figure 5). These two functions were introduced by Popp and Yan to describe the coupling of coherent states [15].

In order to investigate the origins of DL and DL oscillation, whole leaves, homogenates of leaves, isolated chloroplasts and filtered (0.2μm pores) homogenates of corn salad were measured (Figure 6). DL oscillation can only be observed in whole leaves. Homogenates show DL with the same decay time as whole leaves, but without oscillation. Isolated chloroplasts show also DL, which is much shorter than the DL of homogenates and has no oscillation too. Filtered homogenates show no DL at all, although they contain a lot of chlorophylls and other molecular components of chloroplasts. Compared to the original homogenates, they contain no cells, chloroplasts and other cell organelles, which have been filtered out.

The observation that DL oscillation exists only in whole leaves but not in leaf fractions indicates the relationship between DL oscillation and integrity of leaf tissues and cells. Further experiments show how the aging procedure, in which a leaf loses its cellular integrity gradually, affects the DL pattern of a leaf (Figure 7). A fresh leaf shows a strong DL oscillation. During the aging procedure, its DL intensity even increases, but its DL oscillation diminishes gradually and disappears after three weeks.

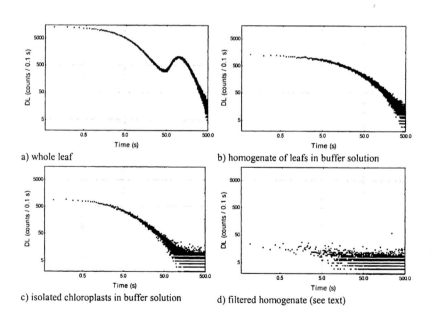

Figure 6. DL-patterns of a whole leaf and leaf fractions of corn salad.

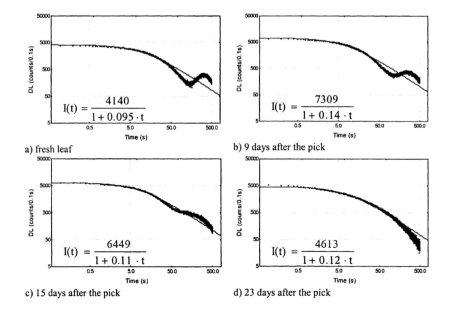

Figure 7. Changes of the DL pattern of an *Anturium* leaf picked from the plant and stored at room temperature.

The above results show that DL of plant leaves is not a phenomenon of molecules, but a phenomenon of tissues, cells and cell organelles. Single light absorbers in plant leaves, e.g. chlorophylls, cannot keep photons for such a long time (>10 minutes). Photon storage is only possible in a complex and ordered system with a large number of light absorbers. The hypothesis that DL origins from PS I and PS II can explain the DL of chloroplasts, but can hardly explain why a whole leaf has an oscillation pattern in its DL decay curve while a solution of chloroplasts displays no oscillation. The DL oscillation of a whole leaf can be explained by couplings of the light sources and sinks in a whole biological system. A coherent system shows such couplings. Theoretically, two coupled coherent states with different damping frequencies lead to a hyperbolic relaxation with an oscillation pattern. The DL decay curve of a whole leaf can be well fitted by the functions describing this kind of coupled coherent states.

4. CONCLUSION

Both BPE and DL contain important information about a plant and its cells. BPE intensity is correlated to the activities of cells, e.g. cell growth and respiration. It is useful in monitoring biological procedures like germination and development. The advantage of a BPE measurement is that it allows a continuous and non-invasive record of a whole biological process. In such a measurement, even small changes, which are often neglected by biochemical methods, can be detected easily. For example, it is easily to find out the circadian rhythm of the growth of barley seedlings through the rhythmic pattern of their BPE dynamics. However, the information gained from BPE is still limited by noise of the nowadays available photon detectors. Further efforts should be done in the development of new photon detectors with much lower noise. DL is certainly correlated to structural states of a plant and its cells. The difference in the DL decay patterns between a whole leaf and leaf fractions indicates that the DL emitters in a whole leaf are coupled. The DL oscillation in a whole leaf is the result of such couplings. They are based on the high-ordered structures in plant tissues and cells. The homogenization destroys these structures and consequently this coupling. The coupling of photon emissions can be easily explained through the model of a coherent system. This model provides also a profound understanding of living systems. DL dynamics, especially DL oscillation patterns can be useful indicators of states, e.g. freshness and health state of plants as well as plant tissues and cells.

5. REFERENCE

1. L. Colli and U. Facchini, Light emission by germinating plants, *Nuovo Cimento* **12**, 150-155 (1954).
2. B.L. Strehler and W. Arnold, Light production by green plants, *J. Gen. Physiol.* **34**, 809-820 (1951).
3. W. B. Chwirot, Ultraweak photon emission and anther meiotic cycle in Larix europaea (experimental investigation of Nagl and Popp's electromagnetic model of differentiation), *Experientia* **44**, 594 (1988).
4. W. P. Mei, *Ultraschwache Photonenemission bei synchronisierten Hefezellen (Saccharomyces cerevisiae) in Abhängigkeit vom Zellteilungszyklus* (Dissertation, Universität Hannover, 1991).
5. F. A. Popp, K. H. Li, and Q. Gu: *Recent Advances in Biophoton Research and its Applications*, (World Scientific, Singapore, 1992).
6. M. Kobayashi, B. Devaraj, M. Usa, Y. Tanno, M. Takeda, and H. Inaba, Two-dimensional imaging of ultraweak photon emission from germination soybean seedlings with a highly sensitive CCD camera, *Photochemistry and Photobiology* **65**, 535-537 (1997).
7. C. Bao, A study on ultra-weak photon-counting imaging of rice seeds at the stage of germination, *Acta Biophyisca Sinica* **14**, 772 (1998).
8. W. Chen, D. Xing, J. Wang and Y. He, Rapid determination of rice seed vigour by spontaneous chemiluminescence and singlet oxygen generation during early imbibition, *Luminescence* **18**, 19-24 (2003).
9. È. Hideg and H. Inaba, Biophoton emission (ultraweak photoemission) from dark adapted spinach chloroplasts, *Photochemistry and Photobiology* **53**, 137-142 (1991)
10. W. F. Bertsch and J. R. Azzi, A relative maximum in the decay of long-term delayed light emission from the photosynthetic apparatus, *Biochim Biophys Acta* **94**, 15-26 (1965)
11. W. Schmidt and H. Senger, Long-term delayed luminescence in Scenedesmus obliquus I. Spectral and kinetic properties, *Biochimica et Biophysica Acta* **890**, 15-22 (1987)
12. È. Hideg, M. Kobayashi and H. Inaba, The far red induced slow component of delayed light from chloroplasts is emitted from Photosystem II, *Photosynthesis Research* **29**, 107-112 (1991)
13. A. Scordino, A. Triglia and F. Musumeci, Analogous features of delayed luminescence from Acetabularia acetabulum and some solid state systems, *Journal of Photochemistry and Photobiology B: Biology* **56**, 181-186 (2000)
14. F.A. Popp, B. Ruth, W. Bahr, J. Böhm, G. Grass, G. Grolig, M. Rattemeyer, H. G. Schmidt, P. Wulle, Emission of visible and ultraviolet radiation by active biological systems, *Collective Phenomena* **3**, 187-214 (1981)
15. F. A. Popp and Y. Yan, Delayed luminescence of biological systems in terms of coherent states, *Physics Letters A* **293**, 93-97 (2002)
16. B. Chwirot, R. S. Dygdala and S. Chwirot, Quasi-monochromatic-light-induced photon emission from microsporocytes of larch showing oscillation decay behaviour predicted by an electromagnetic model of differentiation, *Cytobios* **47**, 137-146 (1986)
17. D.V. Parkhomtchouk and M. Yamamoto, Super-Delayed Luminescence from Biological Tissues, *Journal of International Society of Life Information Science* **18**, 413-417 (2000)
18. J. D. Bewley, Seed germination and dormancy, *The Plant Cell* **9**, 1055 (1997)
19. G. Richter, *Stoffwechselphysiologie der Pflanzen*, (Thieme, Stuttgart, 1998)

20. Y. Yan, Biophotonenemission von Gerstensamen (*Hordeum vulgare L.*), (Dissertation, University of Mainz, 2002)
21. M. Gruwel, B. Chatson, X. S. Yin, and S. Abrams, A magnetic resonance study of water uptake in whole barley kernels, *International Journal of Food Science and Technology* **36**, 161 (2001)

Chapter 16

BIOPHOTON EMISSION AND DEFENSE SYSTEMS IN PLANTS

Takahiro Makino[1], Kimihiko Kato[1], Hiroyuki Iyozumi[1], and Youichi Aoshima[1]

1. INTRODUCTION

1.1. Overview of the Plant Defense Cascade System

Plants protect themselves against attack from pathogens such as bacteria, fungi, nematodes, and insects through the development of a defense system. Doke[1] discovered an "oxidative burst (OXB)" in the defense system of potato tubers. In general, OXB is the central component of the plant defense system. It is now widely accepted that the key component of OXB is hydrogen peroxide, which can be generated on cell membranes from superoxides through several mechanisms that include the constitutive expression of superoxide dismutase. Elicitors induce a defensive response through activation of the defense signaling cascade system, while some microorganisms activate Nicotianamide Adenin Dinucleotide Phosphate (NAD (P)) H-dependent reductase to reduce molecular oxygen to superoxide, which undergoes dismutation either spontaneously or enzymatically to hydrogen peroxide[2]. Reactive oxygen intermediates have been associated with apoptosis of mammalian cells, indicating a role in cell death during the hypersensitive response in plants[3]. Small amounts of reactive oxygen metabolites act as signals for the induction of the detoxification mechanism that involves superoxide dismutases and the activation of other defense reactions in neighboring cells. Lipoxygenase during the hypersensitive reaction in infected tissues may act as a signal for the induction of the defense metabolism in neighboring living cells. Hydrogen peroxide can act directly by virtue of its toxicity to bacteria or fungi, or it can be used as a substrate by extracellular peroxidase in the oxidation of coniferyl alcohol to initiate the lignification chain reaction[4]. Thus, a plant defense response may trigger a chain reaction of active oxygen species.

---

[1] Shizuoka Agricultural Experiment Station, 678-1 Tomigaoka, Iwata, Shizuoka, Japan, 438-0803, Tel: 0538-36-1543, fax: 0538-37-8466, e-mail: makino-log@k7.dion.ne.jp,

1.2. Spectral Analyses of the Photon Emission and Change of Physiological State in the Defense Response

Plants generate a relatively high level of biophoton emission as a response to external changes[5] such as anaerobic treatments,[6] growth hormone treatments,[7,8] saline stresses,[9] temperature changes,[10] herbicide treatments,[11] and attack by pathogens[12]. Biophoton emission depends strongly on the physiological state of the living organisms[13].

The sources of biophoton emission have been of considerable interest, and various investigators have examined the possible sources of biophoton emission from plants, in some cases by using a spectral analysis of photon emission[14,15]. As a result, some sources of biophoton emission, which include enzymatic reactions and reactive oxygen species (ROS), have been determined[14]. However, the total mechanism of biophoton emission is still unclear.

The spectrum of biophoton emission from living organisms is affected by both the species and the contribution ratio of emitters that depend naturally on the physiological state of the living organisms[13]. Therefore, the spectrum of biophoton emission is assumed to reflect the physiological state. However, there have been no attempts to continuously observe the process of physiological transition in living organisms by measuring the spectrum; it is important to measure the spectrum continuously over a long time in order to reveal the dynamic nature of the physiological state. From previous investigations of the spectrum of biophoton emission,[6,14] it was deduced that for optimal measurements of the spectrum of biophoton emission from plants, the intensity of the photon emission should be relatively strong and samples should not contain strongly fluorescent compounds, in particular chlorophyll. We previously reported a biophoton emission generated during the interactions between the sweet potato and *Fusarium oxysporum* that was associated with a defense response[12]. Because the intensity of the biophoton emission from the sweet potato showed that the defense response was very strong, and the storage root of the sweet potato does not contain chloroplasts, this system was considered suitable for spectral analysis. We report here information obtained from continuous spectral analyses of the biophoton emission from the sweet potato during the defense response.

1.3. Approach to Possible Sources of Photon Emission in the Defense System

Several possible sources of biophoton emission have been reported. Of these, ROS or organic radical species, which are produced in enzymatic reactions, have been correlated with photon emission, ROS which can be induced by pathogen infection or elicitor treatments, may be the trigger that initiates signal transduction. This is considered an important possible source of photon emission. This signal results in lignin production, phytoalexin production or Pathogenesis Related (PR)gene expression in the plant defense system. In response to this signal transduction, several kinds of enzymatic activities increase rapidly[4]. LOX, POD and NADPH oxidase, which are related to ROS generation or lipid peroxidation.

Nitric oxide (NO), nitric oxide synthase (NOS) and nitrate reductase (NR), which are involved in nitrogen metabolism, also produce the free radical, NO^{16}. Therefore, these enzymes may influence the photon emission associated with the defense response. Here, we analyze the correlation between the enzymatic reactions and photon emission involved in the plant defense system. As an experiment, we used the biological defense strategies of the sweet potato in response to treatment with the elicitor substrate "chitosan". The intensity of biophoton emission was very strong from the sweet potato[12] when showing the defense response; the latter was clearly detectable by the production of phytoalexin (ipomoeamaron). To identify the enzymatic reactions closely associated with photon emission from plant tissues, we reproduced each enzymatic reaction *in vitro* and examined the correlation between the enzyme reaction and the intensity of photon emission. Our data showed that POD, and especially LOX, were closely associated with the photon emission in a plant defense system.

2. MATERIALS AND METHODS

2.1. A Photon Counting System

A Multi-Sample Photon Counting (MSPC) System II and PCX 100 (Hamamatsu Photonics K.K., Hamamatsu, Japan) were used to observe time-dependent intensity variations in several spectral regions. For example, the MSPC System II is equipped with an R329P photomultiplier tube (PMT) that provides a spectral response from 240–630 nm and a special dark box system with two rotating disks: one disk for 16 samples and another for band-pass filters (Figure 1). The disk with band-pass filters rotates between the PMT and the samples to insert the filters between them. Seven band-pass filters (ca 50 nm pass band each) were used to cover the spectral region from 280–630 nm. The intensity of emission from the blank is 20–30 counts per second (cps) without cooling.

2.2. Inoculation of Microorganisms and the Treatment of Reagents

Radish (*Raphanus sativus* var. *macropolus* cv. Ao-kubi) root and sweet potato (*Ipomoea batatas* var. *batatas* cv. Beni-azuma) storage root samples were obtained by carving cylindrical sections (46 mm in diameter; 6–8 mm thick) from the roots. The disks of radish and sweet potato were placed in polystyrene Petri dishes (60 mm in diameter). Conidia of a non-pathogenic *F. oxysporum* (NPF) isolate were prepared as described previously[12]. A 0.2 ml portion of spore (conidial) suspension (10^7 conidia/ml) or distilled water was applied uniformly to the upper surface of the radish and sweet potato disks. A 0.2 ml portion of 4-(5,6-dimethoxy- benzothiazolyl) phthalhydrazide (DBPH) from Dojin Chemicals, Kumamoto Japan, was applied as a sensitizer for hydrogen peroxide and ROS that were diluted to a final concentration of 10 mM at pH 7. Two disks were used for each treatment. Immediately after inoculation, biophoton emission from the treated surface of the disks was measured at 20 °C for a maximum of 40 h. Each sample was measured every 8 min for 40 s (5 s for filter-less measurements and for the measurements with each band-pass filter). The experiment was repeated three times.

2.3. Application of 2, 4-Dichlorophenoxyacetic Acid

2, 4-Dichlorophenoxyacetic acid (2, 4-D) is a synthetic auxin that promotes cell extension and controls their differentiation. At the appropriate concentration, 2, 4-D induces the sweet potato to form embryogenic calli. 2, 4-D was purchased from the Wako Pure Chemical Industry, Osaka, Japan, and a solution of 2 mg/l was prepared in 10 mM Tris–HCl buffer (pH 7.0). A 0.2 ml portion of the 2, 4-D solution was applied to sweet potato disks. Biophoton emission from the treated surface was measured at 20 °C for

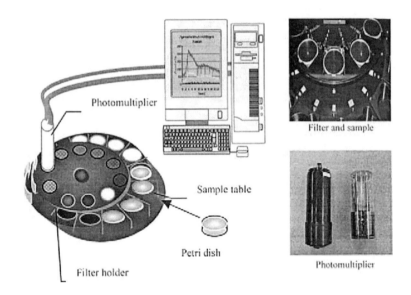

Figure 1. Configurations of the Multi-Sample Photon Counting System (MSPC) II.

65 h after treatment. Each sample was measured using the same method as the samples treated with *F. oxysporum*.

2.4. Control of Sample Temperature

Before long-term storage, sweet potato is incubated for one week at high humidity and high temperature to increase storage quality (e.g. a relative humidity of 90–95%, 30–32 °C). The respiration rate increases and cork layers develop on wounded surfaces during this process. The temperature in the sample chamber was kept at 20 °C for 4 h, increased to 33 °C for 10 h and then decreased to 20 °C for 10 h. The accuracy of the temperature control was ± 0.1 °C. Each sample was measured using the same method as the samples treated with *F. oxysporum*. During the measurements, the temperature of the sweet potato disks was monitored by thermocouple thermometers (TR-52; T&D Co., Matsumoto, Japan) with thin probes inserted directly into the samples.

2.5. Spectral Analyses

The luminescence intensity I_i within each pass band is expressed by:

$$I_i = (N_i - N_{ib}) / (QE_i \times T_i),$$

where N_i and N_{ib} are the number of photons observed through the ith filter from a sample and from a blank holder, respectively, QE_i is the average value of the quantum efficiency of the PMT that corresponds to the pass band of the ith filter, and T_i is the average transmittance value of the *i*th filter. The spectrum of biophoton emission from the sweet potato samples was represented as the ratio of intensity in each spectral region to the integrated intensity of all the spectral regions. The spectral transition was plotted using a moving average of 10 measurements. Only four spectral regions of 430–480 nm, 480–530 nm, 530–580 nm, and 580–630 nm were shown in the Figures because almost no emission was observed in the other three regions.

2.6. Photon Emission and Enzyme Reactions *In Vitro*

We examined about eight enzymes known to act in plant defense systems. Each enzyme and its corresponding substrate were dissolved in 10 mM phosphate buffer (pH 5.2–7.5) as the reaction mixture. Mixtures (5 ml) were placed into polystyrene Petri dishes (60 mm in diameter) and two sweet potato disks were used for each treatment. The time-dependent intensity of biophoton emission from each sample was measured every 114.7 s at 25 °C for up to 13 h and the experiment repeated three times. We used, in preference, materials also involved in living systems as the substrates for enzyme reactions. To monitor the performance of the reactions, we estimated the reaction product for each sample after the photon emission was measured.

The enzymes examined (and their corresponding substrates) were phenylalanine ammonia lyase (phenylalanine), POD (coniferyl alcohol), LOX (linoleic acid), chitinase (ethylene glycol chitin), glucanase (ethylene glycol chitosan), NADPH oxidase (NADPH), NOS (l-arginine) and NR (sodium nitrite).

2.7. Photon Emission and Enzyme Reactions *In Vivo*

POD and LOX induced high levels of biophoton emission in the *in vitro* reaction system. To examine their contribution to the photon emission from plant tissues, we analyzed the influence of enzyme solutions, substrate solutions, and enzyme inhibitors on photon emission.

2.7.1. Application of the Enzyme Solutions

Disks of sweet potato were placed in polystyrene Petri dishes. Solutions of phospholipase (PLA; 10 mg/ml), LOX (10 mg/ml), LOX (10 mg/ml) + PLA (10 mg/ml), or POD (0.05 mg/ml) were prepared in 10 mM Tris-HCl buffer (pH 6.0). A 0.3 ml aliquot of each enzyme solution was applied uniformly to the upper surface of a sweet potato disk and the photon emission from each sample measured.

2.7.2. Application of the Substrate Solutions

Solutions of 2 mM linoleic acid or 0.2 mM coniferyl alcohol were prepared in 10 mM Tris-HCl buffer (pH 6.0) as substrates for the LOX and POD reactions, respectively. A chitosan solution (0.2%) was applied to the upper surfaces of the sweet potato disks. Immediately after treatment, the photon emission from the treated surfaces of the disks was measured at 23 °C. The corresponding substrate solution was applied to the surfaces of the disks when the photon emission had just started to increase (4 h after treatment with chitosan). The photon emission from each sample was measured continuously.

2.7.3. Application of the Enzyme-Reaction Inhibitors

Using the method described above, a solution of 1 mM caffeic acid was used as an inhibitor of the LOX reaction and 50 µM sodium metavanadate as an inhibitor of the POD reaction. These inhibitors were applied to the surfaces of sweet potato disks about 5–7 h after treatment with chitosan. To check the effects of the enzyme inhibitors, we measured the enzyme activity in the inhibitor-treated samples when the photon emission reached its highest peak (about 14–18 h after treatment with chitosan). The sample was then homogenized with 0.1 M sodium phosphate buffer (pH 6.0) and after centrifugation the supernatant was fractionated, LOX and POD assays were performed spectrophotometrically.

3. RESULTS

3.1. Time-Dependent Analyses of Biophoton Emission

Photon emission from radish root slices showed two peaks after activation of the defense system inoculated with NPF. The first peak appeared several minutes after inoculation of the NPF. An ageing treatment, where the slices were left in a high humidity condition overnight, quickened the response (data not shown). The second peak occurred 3–4 h after inoculation and showed a higher biophoton emission intensity than the first peak (Figure 2a). The first peak of photon emission was greatly enhanced with hydrogen peroxide and active oxygen species sensitizer DBPH. However, the second peak was not enhanced under the same treatment (Figure 2b). Compared with the NPF inoculated positive control, the second peak of photon emission was greatly suppressed with

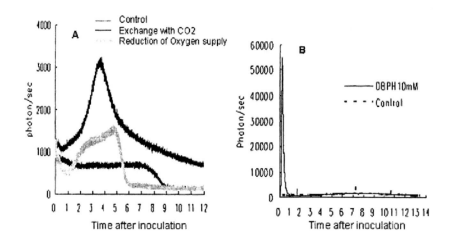

Figure 2. (A) Photon emission from radish inoculated with NPF (positive control) and with a suppressed oxygen supply applied by sealing the Petri dish with film or replacing the oxygen with carbon dioxide. (B) Effect of photon emission enhanced with DBPH (10 mM).

a limitation of the oxygen supply by sealing the Petri dish and replacing the air with carbon dioxide.

3.2. Spectral Analyses of Biophoton Emission

Sweet potato disks treated with water showed no noticeable large variations in the intensity of biophoton emission. In contrast, those inoculated with NPF showed a transient increase in biophoton emission (Figure 3a). The biophoton emission from sweet potato treated with water consisted almost entirely of emission in the three regions of 480–530 nm, 530–580 nm and 580–630 nm (Figure 3c). A small oscillatory variation was observed in about 3 h intervals in the 580–630 nm and 530–580 nm regions. The oscillations between the two regions showed an inverted reflection in a steady rhythmic motion (Figure 3c).

No detectable rhythmic oscillations were observed in the other two regions. Although the spectrum was constant, when the sweet potato was inoculated with NPF the spectrum changed dramatically and the total emission intensity increased (Figure 3b). In addition, emission in the 430–480 nm region was added as a major component of the total emission. While the ratio of the emission in the 580–630 nm region fell sharply, the ratio in the 480–530 nm region showed a sharp rise and those in the 430–480 nm and 480–530 nm regions increased slightly. These changes occurred from 2–10 h after inoculation, during which time the intensity increased slowly. The spectrum was rather stable from 12 h to the end of the measurements, but the intensity kept increasing until 20 h (the peak). From this transition behavior of the spectrum and intensity, the emitters of biophoton emission in the sweet potato change as follows. The species or contribution ratio of the emitters changes greatly from 2–10 h after inoculation, although the quantity of emitters does not increase greatly during this time. From 12 h after inoculation, the species and contribution ratio of emitters are rather constant, but the quantity of emitters increases until 20 h. The oscillations of the two regions disappeared in reversed rhythmic motions after 2 h to the end of the measurements after the inoculation with NPF with magnified observation. However, the oscillations were rather synchronized in the phases between the 580–630 nm and 480–530 nm regions 2 h after inoculation.

3.3. Comparative Analyses with 2, 4-D and Alternating Temperature Treatments

A transient increase of the total emission reached a peak approximately 32 h after the application of 2,4-D (Figure 4a). Increases of the ratios in the 480–530 nm and 530–580 nm regions and a decrease in the 580–630 nm region were observed soon after the application of 2,4-D and the spectrum became quite stable from 16 h after application. The total emission in this stable state consisted almost entirely of emission in the four regions of 430–480 nm, 480–530 nm, 530–580 nm, and 580–630 nm (Figure 4b). Small oscillatory variations were observed in the same regions in the 580–630 nm and 530–580 nm regions. The rhythmic oscillations in the two regions were greatly disturbed and first shortened their intervals 20 h after treatment. However, the rhythmic oscillations appeared 20 h after the 2,4-D treatment (Figure 4b).

With a change in the ambient temperature, the sample temperature increased rapidly from 20–33 °C in 2 h, where it stayed for approximately 8 h before decreasing to 20 °C in 4 h (Figure 4c). The total emission intensity from the sweet potato increased in parallel with the increase in sample temperature and both parameters reached their maximums

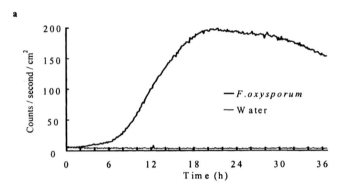
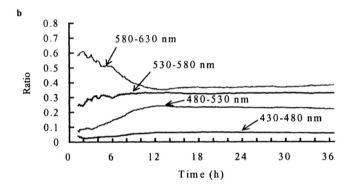
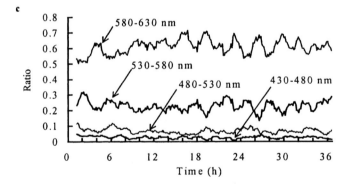

Figure 3. Biophoton emission from sweet potato treated with NPF and distilled water. (a) Intensity of the biophoton emission. (b) Spectral transition following inculation with a conidial suspension of NPF. (c) Spectral transition in the water treatment. Biophoton emission was measured at 20 °C with the MSPC II. A moving average of 10 measured values was used in (a) and (b).

simultaneously. Following this, although the total emission intensity started to drop, the sample temperature remained constant for approximately 8 h. As shown in Figure 4c, d, the spectrum of biophoton emission changed with an increase in the total emission intensity.

The ratio of emission intensity in the 580–630 nm region rose, whereas the ratio in the 530–580 nm region dropped. During a drop in the total intensity at a constant sample temperature of 33 °C, the spectral distribution remained unchanged. However, during the cooling process, the spectrum changed with a fall in total intensity; the ratio of emission intensity in the 580–630 nm region showed a sharp decrease, whereas the ratio in the 530–580 nm region increased. The rhythmic motion of the oscillations in the 580–630 nm and 530–580 nm regions were greatly disturbed and their intervals shortened. However, the inverted reflections of the two regions remained constant during the measurements, regardless of ambient temperature changes.

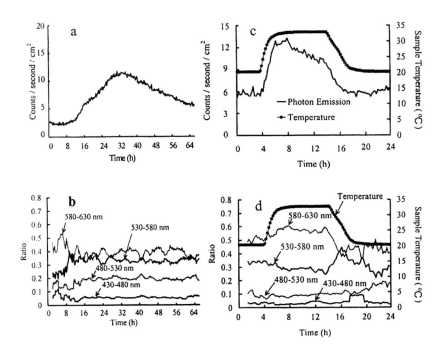

Figure 4. The intensity (a) and spectral transition of biophoton emission from sweet potato treated with 2,4-D. The sweet potato was treated with a 2-mg/l solution of 2,4-D. The intensity (c) and spectral transmission (d) of biophoton emission from sweet potato exposed to alternating temperatures. A sample temperature is shown in (c) and (d). Biophoton emission was measured at 20 °C with the MSPC II. A moving average of 10 measured values was used in (b) and (d).

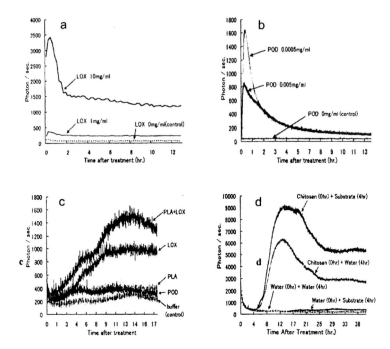

Figure 5. (a) Photon emission from a 5 ml solution containing LOX (0, 1, and 10 mg/ml) and 2 mM linoleic acid at 25 °C. (b) Photon emission from a 5 ml solution containing POD (0, 0.0005, and 0.005 mg/ml) and 0.2 mM coniferyl alcohol at 25 °C. (c) Photon emission from sweet potato slices after the application of 2 mM linoleic acid (substrate) or water at 25 °C, 4 h after treatment with chitosan or water. (d) Photon emission from sweet potato slices after the application of 2mM linoleic acid (substrate) or water at 25 °C, 4h after treatment with chitosan or water.

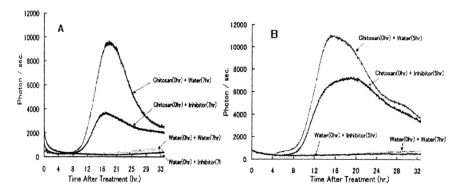

Figure 6 (A) Photons emitted from sweet potato slices following the application of 1 mM caffeic acid (inhibitor) or water at 25 °C, 7 h after treatment with chitosan or water. (B) Photons emitted from sweet potato slices following the application of 50 μM sodium metavanadate (inhibitor) or water at 25 °C, 5 h after treatment with chitosan or water.

3.4. Photon Emission and Enzyme Reactions *In Vitro*

Interestingly, the photon emission was detected only from the POD and LOX reactions in all of the enzymes analyzed. The peak of photon emission was at about 1 min (LOX) or 30 min (POD) after treatment with the enzymes (Figure 5a, b).However there was no detectable emission from the other enzymatic reactions (data not shown). By monitoring the reaction products, we confirmed the absolute performance of each enzyme reaction (data not shown).

3.5. Photon Emission and Enzyme Reactions *In Vivo*

3.5.1. Application of the Enzyme Solutions

Strong stable photon emissions were detected from sweet potato slices treated with LOX. The emission from the samples treated with PLA+LOX was considerably stronger than that from the samples treated with LOX only (Figure 5c). On the other hand, no noticeable photon emission was detected from the POD reaction system.

3.5.2. Application of the Substrate Solutions

The intensity of photon emission from chitosan-treated sweet potato samples increased transiently with the application of linoleic acid, the substrate of the LOX reaction (Figure 5d). On the other hand, the application of coniferyl alcohol, the substrate of the POD reaction, had no obvious effect on the intensity of photon emission (data not shown)

3.5.3. Application of the Enzyme-Reaction Inhibitors

The application of caffeic acid greatly suppressed photon emission from chitosan-treated sweet potato from about 9,500 cps to 3,500 cps (Figure 6a). The application of sodium metavanadate also suppressed photon emission, from about 11,000 cps to 7,500 cps (Figure 6b), although less effectively than caffeic acid. Furthermore, the application of enzyme inhibitors abolished POD and LOX activity when the transient increase in photon emission reached its peak (data not shown). Therefore, both enzyme reactions were completely inhibited by their respective inhibitors.

4. DISCUSSION

Plants have been protecting themselves from attacks by pathogens such as bacteria, fungi, nematodes, and insects through constant evolution in the natural world. Invasion of pathogenic microorganism or non-pathogenic *F. oxysporum* activate the defense cascade system in plant cells. The defense cascade system produces many excited molecules by way of OXB due to enzymatic signaling processes in the cells[1,2]. It is possible for excited molecules to emit photons. We investigated time-dependent analyses of photon emission, their spectral analyses, and then examined the possible sources of photon emission in the plant defense system. Radish root-NPF inoculated samples showed two peaks in photon emission. The first peak appeared a few minutes after inoculation and was dramatically enhanced by chemiluminescence reagent DBPH catalytic hydrogen peroxide and active oxygen species. OXB is a physiological change that is rapid, transient, produces a huge amount of ROS, and is the earliest feature of the plant defense system[2]. The behavior of the plant during the first peak of photon emission and during the OXB phase was approximately the same. However, ROS reagent DBPH did not enhance the secondary peak of photon emission in the NPF inoculated radish samples. These two peaks of photon emission are considered to be controlled by different mechanisms. Plants consume large amounts of energy in maintaining their defense systems and the energy is supplied by the respiratory system[17]. For this reason, we investigated the effect of a restriction in oxygen supply. The second peak was suppressed with a restriction of the oxygen supply by sealing the Petri dishes. Moreover, when the air was replaced with carbon dioxide, both the first and second peaks were remarkably suppressed in photon emission. We presumed that the second peak of photon emission was due to related respiratory chain reactions that generated energy for the defense cascade system to produce defense-related substances that include phytoalexins and pathogenesis-related proteins.

Plants consume considerable amounts of energy in synthesizing defense substances[17]. Based on this knowledge, the strongest physiological change is considered to take place when defense-related substances are first produced in plants. Therefore, it is a reasonable assumption that sweet potato, in responding to infection with NPF, starts to produce defense-related substances 10–12 h after inoculation. This is because the spectral transition reflecting the physiological change was completed at that time. However, a biochemical analysis of the synthetic pathway of the defense-related substances is required to demonstrate this assumption. A continuous analysis of the spectrum enabled us to reveal the change of the physiological state in response to external excitation. This analysis is also a non-invasive and real-time method and can be used as a new parameter for identifying the physiological state of an organism[18].

Untreated sweet potato samples emitted photons with rhythmical oscillations at approximately 3 h intervals, especially between the 580–630 nm and 530–580 nm regions, which had clearly inverted reflective motions. However, the NPF inoculated samples not only had greatly disturbed rhythmical oscillations, but also inverted reflective motions in both regions. A similar phenomenon was observed in the 2,4-D treated samples, although the rhythmical oscillations and inverted reflective motions occurred 24 h after treatment. On the other hand, although the biophoton emission of the temperature treatment samples had disturbed rhythmical oscillations, their reversed reflected motions were maintained between the two regions. Plants have a natural healing system and the sliced sweet potato spontaneously recovered by suberizing the surface layer with lignin using the enzymatic chain reaction mentioned above. The metabolic cost of making lignin is very high[17] and its biosynthesis must be carefully regulated to balance the demands of surface recovery against the availability of resources in making the compound. Therefore, the rhythmical oscillations may be important in optimizing the enzymatic chain reactions.

Next, we investigated the source of photon emission according to physiological changes with the plant defense response. ROS can be induced by pathogen attacks or elicitor treatments and could be the trigger that initiates signal transduction, which is considered a possible source of photon emission. We verified the photon emission of major enzymes expressed in the possible pathway of the plant defense cascade system under *in vitro* and *in vivo* conditions[22].

LOX and POD reactions caused photon emission under *in vitro* conditions (Figure 5a, b). Both enzyme activities are known to increase rapidly in response to certain stimuli. The LOX reaction is directly involved in endogenous lipid hydroperoxidation and its activity rapidly increases in response to elicitor treatment or pathogen infection. The LOX reaction is associated with the early stage of the plant defense system[19]. The POD reaction causes the generation of ROS and the polymerization of phenyl propanoid radicals into lignin in cell walls[20, 21].

We analyzed the influence of these enzymes on the photon emission associated with the plant defense system. The LOX reaction was shown to produce photon emission directly (Figure 5c) and the intensity of photon emission was higher from sweet potato treated with PLA+LOX than from samples only treated with LOX. PLA provides linoleic or linolenic acids, the substrates for LOX that resolve membrane lipids. Therefore, the LOX reaction on PLA-derived substrates may affect the increase in intensity of photon emission from the sweet potato. Furthermore, LOX is associated more closely with photon emission than POD. The application of linoleic acid, a LOX substrate, brought about an increase in photon emission associated with the chitosan treatment (Figure 5d). Before the application of substrates, LOX activity is thought to become higher in the signal transduction process in response to chitosan treatment. The LOX reaction would then be maintained by the supply of linoleic acid, which may affect photon emission.

The photon emission from chitosan-treated sweet potato was apparently suppressed by the application of POD or LOX inhibitors (data not shown). Therefore, both enzymes are considered associated with photon emission. However, the intensity of photon emission was decreased more significantly by LOX inhibitors than by POD inhibitors. Therefore, LOX would seem to be more important in photon emission. How downstream reactions are related to photon emission in the details of underlying the decrease in photon emission by interrupting the LOX reaction is a matter of conjecture. We anticipate that the sources of biophoton emission will be analyzed in detail in future with gain- and loss-of-function studies using transgenic plants as a biological approach. As a physical approach, the biophoton emission will need quantization

5. SUMMARY

A time-dependent analysis of biophoton emission from the radish-*Fusarium* system showed the event of the defense response to the invasive microorganism. We observed two peaks of photon emission in the system. Using a chemiluminescence probe, the first peak was found to appear several minutes after inoculation and the intensity of photon emission was greatly amplified for hydrogen peroxide and active oxygen species associated with the cell membrane. However, the second peak was not amplified under the same treatment and the two peaks of photon emission were suggested to be controlled by different mechanisms. A spectrometry of biophoton emission from the sweet potato showed the defense response was observed during the process of physiological transition. The spectrum showed a drastic transition from 2–10 h after stimulation of the defense system following inoculation with the microorganism. Although the spectrum was stable from 10–36 h after inoculation, the emission intensity peaked approximately 20 h after inoculation. A change in the physiological state connected with the synthesis of defense-related substances is suggested as contributing to this phenomenon. The possible changes in the physiological state connected with lipoxygenase (LOX) and peroxidase (POD) reactions, which are involved in the production of radical species, are shown to be associated with biophoton emission in plant defense systems. The application of LOX to sweet potato slices caused photon emission directly in plants. The LOX substrate promoted photon emission in chitosan-treated sweet potato and the LOX inhibitor remarkably suppressed this emission. Therefore, a LOX-related pathway, which includes LOX and other downstream reactions, is principally associated with photon emission in plant defense systems.

6. REFERENCES

1. N. Doke, Y. Miura, L. M. Sanchez, H. J. Park, T. Noritake, H. Yoshioka, and K. Kawakita, The oxidative burst protects plants against pathogen attack: mechanism and role as an emergency signal for plant bio-defense. *Gene* **179**, 45–51 (1996).
2. N. Doke, NADPH-dependent O_2-generation in membrane fractions isolated from wounded potato tubers inoculated with *Phytophthora infestans*. *Physiol. Plant Pathol.* **27**, 311–322 (1985).
3. J. T. Greenberg, Programmed cell death: a way of life for plants. *Proc Natl Acad Sci U S A* **93**, 12094–12097 (1996).
4. K. E. Hammond-Kosack, and J. D. G. Jones, Resistance gene-dependent plant defense responses. *The Plant Cell* **8**, 1773–1791 (1996).
5. F. -A. Popp, in: *Biophotonics and Coherent Systems*, edited by L. Beloussov, F.-A. Popp, V. L. Voeikov and R. van Wijk (Moscow University Press. Moscow, Russia, 2000), pp. 117–133.
6. P. Roschger, B. Devaraj, R. Q. Scott, and H. Inaba, Induction of a transient enhancement of low level chemiluminescence in intact leaves by anaerobic treatment. *Photochem. Photobiol.* **56**, 281–284 (1992).
7. S. Kai, T. Ohya, K. Moriya, and T. Fujimoto, Growth control and photon radiation by plant hormones in red bean. *Jpn. J. Appl. Phys.* **34**, 6530–6538 (1995).
8. M. L. Salin, and S. M. Bridges, Chemiluminescence in soybean root tissue: effect of various substrates and inhibitors. *Photobiochem. Photobiophys.* **6**, 57–64 (1983).
9. T. Ohya, H. Kurashige, H. Okabe, and S. Kai, Early detection of salt stress damage by biophoton in red bean seedlings. *Jpn. J. Appl. Phys.* **39**, 3696–3700 (2000).
10. P. Roschger, R. Q. Scott, B. Devaraj, and H. Inaba, Observation of phase transition in intact leaves by intrinsic low-level chemiluminescence. *Photochem. Photobiol.* **57**, 580–583 (1993).
11. E. Hideg, and H. Inaba, Dark adapted leaves of paraquat resistant tobacco plants emit less ultraweak light than susceptible ones. *Biochem. Biophys. Res. Commun.* **178**, 438–443 (1991).
12. T. Makino, K. Kato, H. Iyozumi, H. Honzawa, Y. Tachiiri, and M. Hiramatsu, Ultraweak luminescence generated by sweet potato and Fusarium oxysporum interactions associated with a defense response. *Photochem. Photobiol.* **64**, 953–956 (1996).
13. B. W. Chwirot, in: *Biophotons*, edited by J.-J. Chang, J. Fisch and F.-A. Popp (Kluwer Academic Publishers, Ordrecht, The Netherlands, 1998) pp. 229–237.
14. F. R. Abeles, Plant chemiluminescence. *Annu. Rev. Plant Physiol.* **37**, 49–72 (1986).
15. R. Van Wijk, and D. H. J. Schamhart, Regulatory aspects of low intensity photon emission. *Experientia* **44**, 586–593 (1988).
16. J. Durner, D. Wendehenne, and D. F. Kiessig, Defense gene induction in tobacco by nitric oxide, cyclic GMP, and cyclic ADP-ribose. *Proc. Natl Acad. U S A.* **95**, 10328–10333 (1998).
17. T. Akazawa, and I. Uritani, Phytopathological chemistry of black-rotten sweet potato. Part 20. The respiratory increase, phosphate and nitrogen metabolism in the rotten sweet potato. *J Jpn. Soc. Biosci. Biotechnol. Agrochem.* **29**, 381–386 (1955). [in Japanese]
18. H. Iyozumi, K. Kato, and T. Makino, Spectral shift of ultraweak photon emission from sweet potato during a defense response. *Photochem. Photobiol.* **75**, 322–325 (2002).
19. C. Gobel, I. Feussner, A. Schmidt, D. Scheel, J. Sanchez-Serrano, M. Hamberg, and S. Rosahl, Oxylipin profiling reveals the preferential stimulation of the 9-lipoxygenase pathway in elicitor-treated potato cells. *J Biol Chem.* **276**, 6267–6273 (2001).
20. R. Vera-Estrella, E. Blumwald, and V. J. Higgins, Effect of specific elicitors of *Cladosporium fulvum* on tomato suspension cells: Evidence for involvement of active oxygen species. *Plant Physiol.* **99**, 1208–1215 (1992).
21. H. Fukuda, Xylogenesis: Initiation, progression, and cell death. *Annu. Rev. Plant Physiol Plant Mol. Biol.* **47**, 299–325 (1996).
22. Y. Aoshima, K. Kato, and T. Makino, Endogenous enzyme reactions closely related to photon emission in the plant defense response. *Indian J Exp. Biol.* **41**, 494–499 (2003).

INDEX

Adenylate kinase 7, 14
Ahrrenius plot 147,148
Anthurium 197
Apoptosis 26-33, 35-38, 43, 52, 154, 205
Aqueous systems 141, 148, 152, 153
Asymmetric 121
Auxin 207
Bacteria 100, 123, 141, 169, 205, 215
Band-pass filters 207
Barley seed 195-200, 202
Bax 26, 27, 30, 31
Biological rhythms 173
Biophoton 109, 111, 114, 116, 117, 120-140, 144, 154-171, 184, 193, 195, 203, 205-207, 209-213, 216
Biophotonic analysis 141
Biophotonic emission 185, 189, 141
Biophotonic field 141, 142
Biophotonic sources 186
Biophotonics 32, 109, 114, 116, 122, 123, 154, 170, 184, 218
Bio-sensor 32, 33, 37

Blood 53, 54, 61, 62, 70, 71, 73, 74, 77-84, 144, 145, 147-149, 153, 166, 168, 180-182
Blood flow 53, 61, 62, 70, 71, 74, 75, 77-84, 180-182
Brain 59, 73, 75, 82-84, 107, 156, 163-165, 167, 168, 171, 184
Brain slices 164, 165, 167, 168
Caffeic acid 210, 214, 215
Carbon dioxide 210, 211, 216
Cascade 26, 28, 43, 52, 205, 215, 216
Cavity resonator 110
CCD 18, 22, 36, 52, 75, 78, 79, 87, 156, 158-161, 168, 169, 171, 203
Cell death 25-28, 32, 33, 37, 38, 205, 218
Cell growth 121, 122, 160, 202
Cell Transfection 44
Cerebral blood flow 73-75, 77-81, 83, 84, 182
Chain reactions 149, 216
Chaotic 112-117, 119, 120, 122
Chemiluminescence (CL) 85, 87, 91-94, 96, 97, 107, 141, 144, 164, 166-168, 171, 184, 203, 215, 217, 218

220 INDEX

Chitinase 209
Chitosan 206, 209, 210, 214, 215, 217, 218
Chlorophyll 206
Chloroplast 193
Coherent 2, 44, 53, 55, 57, 58, 64, 78, 82, 87, 109-119, 121-123, 127, 134, 140, 141, 154, 190, 195, 201-203, 218
Coherent state 111-113, 118, 126, 127, 154, 190
Composite state 136
Confocal laser scanning microscope 44, 45
Correlation functions 76, 110, 114, 120, 122
Crassula ovata 197, 200
Critical phenomena 149
Cyan fluorescent protein (CFP) 11, 27, 31, 33-36, 41, 42, 44-51
Cytochrome c 26-29, 31, 38, 142
Cytoplasm 7, 8, 19, 21, 27, 44, 45, 102
Defense 69, 183, 205-207, 213, 215-218,
Delayed luminescence (DL) 109, 119-122, 140, 169, 174, 175, 180-182, 187-189, 191, 193-197, 199-203
Detoxification 205
Differentiation 43, 121, 122, 186, 187, 191, 193, 203, 207
Dismutase 86, 96, 143, 205
Dismutation 182, 205
DL decay curve 195, 202
DL oscillation 201-203
DNA 1, 25, 40, 44, 121, 123, 136, 141, 186, 191
Drug screening 37, 54
Electron excitation energy (EEE) 141, 143-145, 148, 150, 152, 153
Electronically excited states (EES) 141, 144, 159, 182
Elicitor 206, 216-218
Embryogenic 207
Emission spectrum 159, 166, 167
Energy transfer efficiency 11, 40, 42
Enzyme 27, 40, 106, 143, 207, 209, 210, 215

Ethydium bromide 1
Ethylene glycol chitin 209
Femto-seconds pulsed laser 44, 45
Fibroblastic differentiation 187, 191, 193
Fluctuation Correlation Spectroscopy (FCS) 1, 2, 14
Fluorescence lifetime 42-44, 47, 49, 51, 52
Fluorescence Lifetime Imaging Microscopy (FLIM) 43, 46, 47, 51, 52
Fluorescence resonance energy transfer (FRET) 27, 31, 38-40, 42-47, 49, 51, 52
FRET microscopy 41, 42
Free radicals 17, 85, 91, 93, 142, 149, 152, 154, 171, 173, 182
Fungi 205, 215
Fusarium oxysporum 206, 218
Fusion proteins 40, 42, 46-50
Gaussian-Lorentzian 2, 4
Green fluorescent protein (GFP) 5-7, 11, 19-21, 25-31, 38, 40, 51, 52
Genetic code 140
Geometrical distribution 113, 115
Germination 14, 161, 197, 199, 202, 203
Glucanase 209
Hamiltonian 119, 126, 127
Hayflick 186
Heat shock protein 27 39, 43, 52
Heat stress 100, 162
HeLa cell 19, 29, 30, 34, 35
Hemoglobin 145
Herbicide 206
Holistic properties 135, 154
Hormone 26, 206
Human body 121, 122, 156, 168, 171, 173, 174, 177, 184
Hydrogen peroxide 47, 154, 186, 205, 207, 210, 215, 217
Hydroperoxidation 216
Hypersensitive 205
ICCD 86, 87, 95

INDEX

Infection 52, 206, 216, 217
Intermediates 205
Kalanchoe daigremontiana 197
Laser Doppler flowmetry 73, 83, 84
Laser speckle contrast imaging 73, 78
Leaf 115, 120, 161, 171, 195-197, 200-203
Lichen 130-133, 140
Light piping 121
Lignification 205
Linoleic 166, 168, 209, 214, 215, 217
Linolenic 217, 166, 168
Lipid peroxidation 160, 16, 169, 184, 186, 206
Lipids 143, 182, 217
Lipoxygenase 205, 217, 218
Low-level photon emission (LLPE) 141, 144, 145
Lucigenin 97, 144, 145
Luminescence 91, 93, 101, 103, 104, 106, 109, 119-122, 140, 144, 154, 155, 169, 171, 174, 175, 178, 180-182, 184, 187-189, 191-195, 203, 218
Luminol 86, 92, 93, 97, 144-148
Maillard reaction (MR) 141, 149-153
Mammalian 33, 100, 160, 185, 187, 188, 191, 192, 205
Mammalian cells 185, 187, 188, 191, 192, 205
MAPK-activated protein kinase-2 (MK2) 44, 49-51
Melanoma cells 187, 191-193
Membrane 2, 4, 7-9, 15, 18, 19, 21-23, 29, 30, 59, 101, 160
Metabolism 73, 82, 83, 95, 142, 159-161, 164, 166, 171, 205, 206, 218
Metabolites 141, 205
Metavanadate 210, 214, 215
Methylglyoxal 149, 151-154
Microorganisms 17, 205, 207
Mitochondria 26-31, 38, 101, 142, 143, 154, 171, 197
Mitogenetic radiation 141, 142, 193
Molecular aggregate 4

Morphogenic field 139
Mouse 44, 87, 90, 94, 96, 154, 156, 163, 165, 187, 191, 192
NAD(P)H 143, 154, 206, 209, 218
NADPH oxidase 143, 206, 209
Nematodes 205, 215
Neutrophils 52, 143-145
Nicotianamide Adenin Dinucleotide Phosphate 205
Non-exponential decay 126
Non-invasive 13, 27, 163, 170, 183, 192, 202, 216
NOS 123, 206, 209
Optical Coherence tomography (OCT) 53-60, 62-71
    frequency and time domain OCT 60
    functional OCT 54, 69
    polarization sensitive OCT 53, 63
    second harmonic OCT 59, 66
Optical Doppler tomography 53, 59, 70, 71
Optical imaging 74, 75, 82
Oscillations 118, 119, 122, 145, 146, 150-153, 181, 191, 211, 213, 216
Oxidative burst 205
Oxidative processes 141, 142, 144, 149, 151, 152
P38 MAP kinase 43, 45-47
Pathogens 143, 205, 215
Peroxidase 161, 162, 205, 217
Phenylalanine ammonia lyase 209
Phosphatidylinositol 3-kinase 44, 49
Photo count statistics 128, 129
Photobleaching 28, 42, 26
Photomultiplier 68, 87, 152, 155, 174, 178, 179, 181, 182, 185-188, 191, 196, 207, 208
Photon counting 5, 6, 9, 43-45, 156-159, 163, 168, 169, 186, 187, 189, 207, 208
Photon counting imaging 156, 168, 169
Photon emission 73, 109, 118-120, 125, 128, 141, 145, 155, 160-169, 173-183, 186, 187, 189-192, 195, 206, 207, 209, 210, 214-218
Photon sucking 118
Photon trapping 115, 118, 119, 121

Photons 4, 43, 54, 91, 112, 113, 116, 119-121,
   125-130, 134, 136, 139, 141-143, 147,
   148, 150, 156, 157, 163, 169, 173, 182,
   185-187, 191, 192, 202, 214-216
Phytoalexin 206, 207
Plant 119, 121, 161, 162, 186, 189, 195, 197, 199,
   200-203, 205-207, 209, 215-218
Poissonian distribution 113-115, 120
Polymerization 217
Probability distribution 75, 129
Programmed cell death 25-28, 37
Protein-protein interaction 27, 39, 40, 41
PS I 195, 202
PS II 195, 202
Pulse-interval counter 157
Quantum 3, 40, 43, 110, 111, 113-116, 121, 122,
   126, 128-131, 134-139, 143, 156
Quantum nature 128, 131, 134, 135, 137, 138
Quantum patch 136, 137
Quantum selection 135, 137
Q-value 131, 132, 134
Radical reactions 113, 117, 120, 141, 162, 168
Raphanus sativus 207
Reactive oxygen species (ROS) 141, 142, 159, 183,
   186, 206
Redox potential 142, 151, 152
Reductase 142, 205, 206
Relative maximum 195
Remote intervention 139
Remote sensing 139
Respiratory burst 144-148, 152
Rhythmic motion 211, 213
Schiff bases 149
Schrödinger question 110

Sensitized Acceptor Fluorescence 42
Sensitizer 85, 207, 210
Signal transduction 37, 41, 206, 216, 217
Single living cell 25-27, 33, 35, 37, 39, 51
Skin 53, 54, 59, 61, 62, 63, 66, 67, 74, 121, 163,
   164, 169, 174-177, 181, 182, 186, 187,
   190, 191
Skin fibroblasts 186, 187, 190, 191
   topographical variation 176
Somatosensory cortex 74, 75, 77, 78, 79, 82-84
Soybean 160, 161, 171, 203, 218
Spatiotemporal analysis 155-157, 162, 163, 165
Spectrum 56, 60, 66, 75, 109, 153, 159, 164, 166,
   167, 173, 174, 178, 186, 188, 206, 209,
   211, 213, 216, 217
Spontaneous biophoton emission 128, 195
Spontaneous emission 173-176, 180-182
Squeezed light 120, 121, 123, 194
Squeezed state 125-135, 138,139
Superoxides 183, 205
Sweet potato 206-218
Temporal Clustering Analysis 72,74
Time series 29, 30, 79,129-134, 151, 175, 176
Time-correlated single photon counting 4, 45
Two-dimensional photon counting tube 156
Two-photon excitation 2, 3, 13, 14, 41, 52
Two-photon fluorescence 2,14
Ultraweak photons 185, 186, 191, 192
Water splitting 152, 154
Wortmannin 49,50,51
Xeroderma pigmentosum (XP) 186
Yellow fluorescent protein (YFP) 11, 27, 51
Zymosan 144-148

Printed by Books on Demand, Germany